AT THE EDGE OF TIME

REALITY, TIME, AND MEANING IN A VIRTUAL EVERYDAY WORLD

DONALD W. JARRELL

February 2013
Revised, October 2014
Copyright: Donald W. Jarrell

CONTENTS

Preface i

Chapter 1. The Basis of Knowing 1

Knowing and Reality 2

Knowing the World Beyond 10
Learning from the Experience of the Mystics 11

Knowing Our Everyday Virtual World 17
Implications of a Virtual World 17
The Pointillist Principle 17
The Information-World Postulate 20
The Logical Imperative 21
What Is Our Everyday World? 22
The Many-Worlds Interpretation of Reality 23
Do Other Worlds Contain Information Useful to Us? 25
Wave Function Collapse and MWI 25
The Delayed-Choice Experiment 27
Can We Retrieve that Information? 30
Transcendent Experiences of Everyday Life 33
Thinking Creatively while in a Self-induced Trancelike State 33
Scientists and the thought experiment. 36
Answers from far and near. 37
Artists, scientists and truth. 40
Thinking Creatively while in a Hypnotically-Induced Trancelike State 42
Performing Creatively while in a Self-Induced Trancelike State 46

Can the Dual-Realities Interpretation Be Proven? 50
Evidence that Information Is Available in Other Worlds 52
Quantum-Level Research 52

Macro-World Research 53
Proof and the "Extra Body" Problem 55

Beyond Evidence; How about Seeking Truth? 56

Does it Matter?: Implications of the Dual-Realities Hypothesis 58

Chapter Summary 61

Questions 62

Chapter 2. At the Edge of Time 69

What Is Time? 71

Continual Creation 75
We Are Moving through Space Not Time 76
A Cosmic Blueprint for Our Lives 77

Personal Time 79
Real Time and Comparative Time 81
Does Time Dilate for Moving Non-Living Things? 84
Comparative Time Differences in Infrequent Circumstances 85
Near Accident or Impending Peril 86
Peak Experience 86
Hypnotic Trance 89
What Happens during Time Distortion? 94
Time Distortion during Impending Peril and Peak Experience 96
Time Distortion during a Hypnotic Trance 98

Can the Holographic Movie Model of Time Be Proven? 100
Continual Creation 100
Continual Creation and the Need for Indivisible Space 100
Each Frame of the Holographic Movie a New Creation 102
Personal Time 103
The SCAD Experiment 104
Concurrent Reporting in Hypnosis As Evidence? 106
Habeas Corpus of a Sort? 106

Chapter Summary.. 109

Questions 110

Chapter 3. Time and Particles in Superposition 115

The Holographic Movie Model of Time and Bell's Theorem 116

The Holographic Movie Model of Time and the Many-Worlds
Interpretation 118

Future Research about Particles in Superposition................................. 121
How We Think Creatively 121
Employing the Particle in Superposition as a Basic Unit of Action....... 122

Chapter Summary.. 124

Chapter 4. Inertia, Gravity, and Time............................. 125

The Linkage of Inertia and Time 126

The Linkage of Inertia, Gravity, and Time 128

Free Fall and Numerical Equivalence... 132

Chapter Summary.. 133

Chapter 5. Life and Meaning in a Virtual World.................. 135

What Is Life and the Universe?.. 137

Is There a Creator? 139

Do Our Lives Have Purpose? 140

Our Primary Purpose in Life: to Choose between Good and Evil 142
The Major World Religions Support This View.............................. 142

We Are by Nature Morally Evaluative 144
Solving the "Problem of Evil" 147
We Are Not Alone 148

Chapter Summary... 150

Chapter 6. Epilogue; Concluding Comments 151

The New Reality Is Better 152

We Can Be Better People 154
A Role for Organized Religion?.. 155

A Speculation: Is There an Afterlife? 158

Bibliography... 159

Index.. 173

About the Author 195

PREFACE

There is nothing so practical as a good theory. (Kurt Lewin)[1]

The field cannot well be seen from within the field. (Ralph Waldo Emerson)[2]

From the outset, fifteen years ago, this was a book about time. But I soon discovered that, to understand time, I had to re-examine

---

[1] D. Cartwright, ed., *Field Theory in Social Science; Selected Theoretical Papers* (New York, NY: Harper & Row, 1951, 169; cited in http://infed.org/mobi/kurt-lewin-groups-experiential-learning-and action-research/.

[2] Ralph Waldo Emerson, *Ralph Waldo Emerson: Selected Essays, Lectures and Poems* (New York, NY: Bantam Books, 1990, 199.

other extraordinarily complex features of our world, notably, the nature of reality itself. This book is the result of that search for understanding.

I am not a physicist. I have a doctorate in an unrelated field from a respected university (University of Pennsylvania) and value properly-done research and experimental method. I am an engineer with credentials (BS, West Virginia University) and experience (Armco Steel Corporation). To me, equally important, I am, by avocation, a critical philosopher of the metaphysicist variety and bring to my thinking a passion for truth. I have tried to be relentlessly logical in my search for truth while also allowing my mind to freely imagine the physical concepts of our everyday life and what might lie beyond what we see each day. This combination, I believe, when combined with an insistence that central concepts of any conceptual framework always be supported by experimental findings, is a requirement for understanding the very special universe we live in. The combination has allowed me to arrive at answers to a number of fundamental questions that satisfy my own very demanding need to know: What is reality? What is time? How will these understandings of reality and time affect the way we understand other basic ideas of physics? I have tried to answer these questions in this book.

Many philosophers and physicists agree that time is only an illusion. It is clear, however, based upon findings of quantum mechanics and neurology, that both time *and* our everyday world are illusions, that what we experience as time and space is a virtual reality. If our everyday world is illusory, why have we not realized this before? The answer I believe is that the laws of the physics of this virtual reality effectively prevent discovery of itself by its empirical sciences. However, while our science cannot penetrate this reality directly, our science can demonstrate that this reality very likely exists. How is this done?

If we believe that we live in a virtual reality, we can test this assumption by asking two questions. First, if we look inside our world at its basic elements—at what our world is made of, are those basic elements intangible as we would expect for a virtual world? And we also might look in the opposite direction, so to speak. If we live in a virtual reality, our virtual reality should appear to emanate from outside itself; does it do so? If the answer to both these questions is "yes", as our science indicates, we must take seriously the likelihood that we live in a virtual world.

If this is a way to demonstrate that we live in a virtual world, why have we not, in all of human history, done this before? Because our science was not sufficiently well-advanced. Only during the lifetimes of humans now living have quantum physics and neurology advanced sufficiently to answer the two questions posed here and thus to show that we almost certainly live in a virtual world.

If we live in a virtual reality there must then be (at least) two realities, our virtual reality and another reality from which our virtual reality emanates. I will speak of this premise as a dual-reality hypothesis, a hypothesis that (as, for example, with the hypotheses of dark energy and dark matter) is impossible to definitively prove or disprove but a hypothesis that satisfactorily explains aspects of our cosmos not otherwise explainable.

While it is not possible to definitively prove or disprove this dual-reality hypothesis, it will self-validate if it is found to be useful.

The ultimate test of any theory is that it serve to explain our world in ways that allow us to understand, to change, and to accurately predict world events. As I will show in the pages below, the dual-reality theory allows us to see, and therefore to understand, our world as we have not seen it before. Just as the view of a field from a mountaintop allows us to see things we would not have noticed when viewing the field from within the field itself, this new view of reality allows us to see our world from a new dimension, "from above", so to speak, in a new, more complete, way. This more complete view allows us to understand for the first time phenomena that previous theories did not allow us to fully understand; these phenomena, presented in Chapters 2, 3, and 4, are:

> the nature of time and why time passes more slowly
>> for some of us (Chapter 2);
> particles in superposition, Bell's theorem, and the
>> many-worlds interpretation of quantum mechan-
>> ics (Chapter 3);
> and the relationships of time, gravity, and inertia
>> (Chapter 4).

Moreover, all of these explanations are themselves subject to testing by experiment. This theory, then, is presented here as a coherent whole subject to testing on a wide variety of fronts.

Chapters 5 and 6 of the book present a sharp change of pace from earlier chapters. The dual-reality theory suggests a much more important role for the individual self in relation to all else than in previous views of reality and time. This shift in the relative importance of the individual demands a new look at questions about individual life and meaning. I offer my views regarding these questions as a first step in the process of what would be, hopefully, a larger conversation.

One feature of the book's structure is unusual and needs explanation—my use of question and answer sets in separate sections at the end of several chapters. These typically concern topics of relevance to the chapter that were part of an earlier draft of the chapter but were edited out in later drafts. While these topics are relevant to material presented in the chapter, they did not fit into the chapter itself as written. I believe my readers will find these questions and answers to be interesting; however, reading these question and answer sets is not necessary for understanding the book's message—the chapter texts convey the essential message of the book.

As a final, very important, prefatory note, I want to acknowledge three debts of gratitude. Without the help of two people this book would not have been written—my wife, Joanne, and my very good friend, Paul Kessler. They are in no way responsible for the shortcomings of the book but the book could not have been written without their unwavering support throughout the writing process. Joanne also listened to my ideas, found helpful publications, and offered a number of excellent suggestions about how the book should be presented to my readers. Paul and I had many helpful discussions of my ideas over lunchtime meals at the Drexel Faculty Club and Paul made a number of helpful suggestions for revision of the first edition of the book. I also want to acknowledge the help of another friend, Keith Wickstrom, for serving as an early reader and for offering suggestions for revisions to the first edition.

CHAPTER 1
THE BASIS OF KNOWING

All that we are arises with our thoughts. With our
thoughts we make the world. (Buddha)[1]

Physical concepts are free creations of the human
mind, and are not, however it may seem, uniquely
determined by the external world. (Albert Einstein)[2]

---

[1] As translated by Thomas Byrom from the *Dhammapada* in Jack Kornfeld, ed.,
Teachings of the Buddha (Boston, MA: Shambhala Publications, 1993), 5.

[2] Albert Einstein and Leopold Infeld, *The Evolution of Physics; The Growth of
Ideas from Early Concepts to Relativity and Quanta* (New York, NY: Simon and
Schuster, 1961), 31.

This book is centered on the question: What is time? To answer this question, however, we must first look at the nature of reality.

KNOWING AND REALITY

We usually assume we know what is real and what is not. Something is real to most of us if we can apprehend it with our senses—if we can see, hear, touch, smell, or taste something it is real. About one hundred years ago, at the turn of the twentieth century, this was the classical view of reality and there was little reason to question this view. However, two things in particular have happened since that time to challenge this line of reasoning and to point the way to a new understanding of reality.

Quantum theory, a branch of physics dealing with physical phenomena at the atomic and subatomic scales, was the first of these events. This theory was set in motion in the early part of the twentieth century when physicists found that the classical theories of the physics of our everyday world could not be used to describe the behavior of particles on the atomic and subatomic scale. The way we thought the world behaved was quite different from the way elementary particles were found to behave. Characteristics of elementary particles that challenged the classical theories—and the classical definition of reality—included:

Quantization of energy: energy is made of fundamental units, indivisible packets, rather than being divisible to a vanishing point;

The uncertainty principle: It is impossible to measure with complete accuracy a particle's position and velocity at the same time—the more accurately you measure one, the less accurately you can measure the other;[3]

Wave-particle duality: particles may behave either as waves or particles.

Each of these discoveries was counterintuitive to the existing view of reality. Especially challenging was wave-particle duality and the so-called two-slit experiment (also called the double-slit experiment) that first demonstrated this duality. In the classic version of the two-slit experiment, first done more than 200 years ago, light rays passing through two parallel slits displayed characteristic wave behavior (think here of water waves) by interfering with each other, creating a pattern of light and dark patches on a photosensitive screen positioned behind the slits. The patches corresponded to the points on the screen where the peaks and troughs of waves diffracting out from the two slits combined with one another either constructively or destructively. When two or more waves come together, the resulting amplitude (wave height) is the sum of the individual waves. Light patches occur when the crests of two light waves come together while dark patches occur when the crest of one wave meets the trough of another wave. Thus, in the classic version of the experiment light was demonstrated to travel as a wave.

However, other research findings suggested that under some circumstances light consisted of discrete quantized packets. In the twentieth century, physicists re-performed this experiment with

---

[3] Note that this is not a measurement problem but rather a principle problem—you cannot, in principle, measure with complete accuracy a particle's position and velocity at the same time.

low-intensity light to show that this interference pattern was evident even when particles of light (photons) passed through the apparatus *one at a time*.[4] Light, therefore, was shown to exhibit both particle-like and wavelike properties. (These experiments can also be done with other subatomic particles with the same results.)

An even more profound challenge to classical reality, however, occurred when a variation of the two-slit experiment was performed placing detectors at the slits to determine through which slit a particle was passing—this act destroyed the interference pattern on the screen behind the slits. The behavior of photons thus was *changed* depending on whether or not an attempt was made to observe them— that is, they behaved as an elementary particle when particle detectors were used or as a wave when detectors were not used.[5] "

If elementary particles do not have definite material existence when we are not trying to observe them can they be tangible objects? This is not the way we expect tangible objects to behave! This finding, together with other characteristics of elementary particles described above, convinced many quantum physicists that elementary particles are not tangible objects but are *bits of information* only. Paul Davies says, "it is easy to be seduced into believing that there really is a little thing 'out there', like a scaled-down version of a billiard ball, producing the results of the measurements. But this belief does not stand up to scrutiny." According to Davies, the

---

[4] David Deutsch, *The Fabric of Reality: The Science of Parallel Universes—and Its Implications* (New York, NY: The Penguin Press, 1997), 38-45.

[5] For a more technical description of this experiment, see Gary Zukav, *The Dancing Wu Li Masters: An Overview of the New Physics*, hardcover edition (New York, NY: William Morrow and Company, 1979), 299.

elementary particles out of which matter supposedly is composed, are not really elementary at all. They are of a secondary, derivative nature. Rather than providing the concrete 'stuff' from which the world is made, these 'elementary' particles "are actually essentially *abstract* constructions based upon ... 'observation events' or measurement records." (Emphasis Davies).[6]

If elementary particles are not tangible objects, what then are the waves associated with these particles? Waves had previously been thought to be moving descriptions of where their associated particles were and how they were moving. Davies believes that waves must be something more abstract from which some statistical information about other abstractions can be obtained. "The wave function represents not how the system is, but what we know about the system."[7] Waves are assemblages of information about the wave as a system.

Both elementary particles and their associated waves thus appear to be *information only*. The reality observed at the quantum level did not bear out the classical view of reality we believed existed throughout our everyday world.

The second event to challenge the classical understanding of reality was neuroimaging which began to be used widely as a research tool in studies conducted in the 1970s.[8] These studies greatly expanded our understanding of how the human brain functions

---

6 Paul Davies, *The Cosmic Blueprint: New Discoveries in Nature's Creative Ability to Order the Universe* (New York, NY: Simon and Schuster, 1988), 175.

7 Davies, *Cosmic Blueprint*, 171.

8 "History of Neuroimaging," Wikipedia, October 31, 2011.

and made us realize how little we know about this incredibly complex organ.[9] The human brain contains about ten billion neurons (nerve cells). On average, each neuron is connected to other neurons through about ten thousand synapses (junctions between nerve cells, each consisting of a minute gap) across which electrical impulses pass. Neurons are constantly exchanging ions (electrically charged atoms or molecules) with surrounding tissue. Ions of like charge repel each other, and when many ions are pushed out of many neurons at the same time, they can push their neighbors, who push their neighbors, and so on, resulting in a wave.[10] Our mental functions are thought to arise from electrochemical signals traveling along the neurons and passing information signals at the junctions, the synapses.[11]

The electrical charge of these waves allows brain-imaging studies of brain function. For example, this charge can be measured by a voltmeter attached to electrodes at various places on the scalp in a type of brain imaging called electroencephalography (EEG). While EEG enables studying brain function by measuring the brain's electrical activity directly, it has a major limitation in that signals measured on the scalp may not represent the activity in the underlying cortex. Other brain imaging techniques, such as single-

---

[9] See Barrow, John D. *The Constants of Nature: From Alpha to Omega-the Numbers That Encode the Deepest Secrets of the Universe* (New York, NY: Pantheon Books, 2002), 118.

[10] Barrow, *The Constants of Nature*, 117-118.

[11] Danah Zohar, "Creativity and the Quantum Self," in *Creativity*, John Brockman (ed.) (New York, NY: Touchstone, 1993), 202-218, 211.

photon emission computed tomography (SPECT), record changes in blood flow that are indirect markers of brain electrical activity. Radioactive tracers and a scanner are used to record data that a computer uses to construct two- or three-dimensional images of active (as indicated by blood flows) brain regions.[12] A very important such study by Newberg, D'Aquili, and Rause looked at what brain imaging can tell us about the nature of human consciousness.

This study showed that, *regardless of whether the world we experience every day is or is not objectively real,* we become conscious of this world only after completion of a complex assembly process in the brain. They say:

> Nothing enters consciousness whole. *There is no direct, objective experience of reality.* All the things the mind perceives— all thoughts, feelings, hunches, memories, insights, desires, and revelations—have been assembled piece by piece by the processing powers of the brain from the swirl of neural blips, sensory perceptions, and scattered cognitions dwelling in its structures and neural pathways. [13] (Emphasis mine)

If the mind builds its rendition of reality by using the things in our world as models, *why is our world deconstructed and put together again before we experience it as reality?* Why do we not, machine-like, simply sense our world and react? On reflection, this study suggests, consistent with the findings of quantum physics, that *our world cannot*

---

12 "History of Neuroimaging," Wikipedia, October 31, 2011.

13 Andrew Newberg, Eugene D'Aquili, and Vince Rause, *Why God Won't Go Away: Brain Science and the Biology of Belief* (New York, NY: Ballantine Books, 2001), 35-37.

be used as a model for constructing the world we experience because our every-day world has no tangible existence apart from what is perceived by our minds. The everyday world we experience is *all* in our minds. Our astonishingly complex brains and minds—the complexity of which is said to rival or exceed, that of the universe itself[14]—assemble and experience reality without need for a previously existing tangible world.

Let me suggest an alternate view of reality that is a better fit with the information we have about our universe and is consistent both with what quantum physics tells us about our world and the brain imaging study of Newberg, D'Aquili, and Rause: the things that we see in everyday life have no reality apart from what our minds perceive—they are all in what I call, for want of a better term, a "universe of the mind," that is, within the *energies* appearing as waves coursing through our minds.

Having said that our everyday world is all in our minds, I believe that, in another sense, it also is *real*. The final arbiter of whether something is "real" need not be whether it is only in our minds or has an objective existence in its own right. Nor should the final arbiter be whether a reality could be proven with the scientific method, a point to be discussed in a later section of this chapter. While it is correct to say that our everyday world is "intangible" (it has no physical presence), it is not "imaginary"; we do not create this everyday world with our imaginations. To the contrary, our world "happens to us"; we do not in any sense create our everyday world. (Exercising choice as to the paths we will follow in our world normally has a trivial effect on

---

[14] Barrow, *The Constants of Nature*, 118.

the totality of the world we experience.) Further, to deny the reality of our everyday world is to deny the existence of the quintessential part of us that makes us human; to do so is, in effect, to deny our humanity for it is within the context of our everyday world that the concept "humanity" is developed and enacted. Further, the pleasures and pains others and we experience in our daily lives matter to us and therefore are real in that sense. And it is for me beyond belief that this elaborate stage show being enacted in our minds has no purpose. And thus, to me, our everyday world (this world of intangible people and physical objects) is real in the same sense as when you say, "Get real," you mean you want the listener to be serious about a topic of discussion. We sense at a fundamental level that our everyday world is real and must be taken seriously.

If you accept the above portrayal of reality, we have then *two* realities, one located outside our everyday world where the script for our everyday world must originate, and one based in an intangible everyday world of elementary particles—which, as we will see later, is far more complex than the world we see each day. How well does modern-day science understand these realities? While the scientific method has been remarkably successful in helping us to understand our everyday world, is there important information about our every-day world that is beyond the reach of science *as it is currently practiced?*

It is probably fair to say that most scientists believe that there *is* knowledge important to understanding our everyday world that lies beyond the reach of modern-day science. Albert Einstein and his coauthor Leopold Infeld compared our search for knowledge about our universe to trying to understand a watch by observing the outside of the watch only. They felt that any future so-called "theory of everything" based on this outside-only view of the watch, however accurate and complete it might be, likely would not be the same as a description based on a view of both the inside and the outside of the watch.[15] And Paul Davies, after much time spent asking and answering an ever-receding "why" says, "'Ultimate' questions will always lie beyond the scope of empirical science as it is usually defined."[16] In short, the modern scientific view of our world does not fully reveal the way things actually are. (The basic reasons for the limitations of modern-day science will be discussed in later portions of this chapter.)

If we are to find additional information to complement our scientific knowledge of our universe, two obvious places that we might search are the domain beyond time and space and the domain within time and space itself. Surprisingly, science itself has much to tell us about the possibilities of finding additional information in each of these areas. I will refer to these two domains as the "world beyond" and the "everyday world." We now need to examine what we can know about information that may be contained in each of these domains.

Knowing the World Beyond

As suggested above, descriptions of our universe by the scientific community are formed by the process of deriving general principles and premises from our observations in our everyday world. Is

---

[15] Albert Einstein and Leopold Infeld, *The Evolution of Physics,* 31. (Original copyright, 1938).

[16] Paul Davies, *The Mind of God* (New York, NY: Simon and Schuster, 1992), 15.

it possible to see our universe in a different way to complement this scientific view? As discussed earlier in this chapter, there seems to be a script for our daily lives originating from somewhere outside our everyday world. We say this script originates outside our everyday world since logic tells us that the images of our everyday world, our universe, cannot be projected to us by those same images. Is there information in that "world beyond" our world of time and space that might help us to answer some of the questions about our everyday lives that our science cannot answer? The experience of mystics may help to answer these questions.

Learning from the Experience of the Mystics

Attempting to look beyond our everyday world is, of course, what religious scholars, mystics, and ordinary people of faith or no faith have been doing for perhaps at least as long as humankind has been using inductive reasoning: they engage in spiritual activity in an attempt to go beyond the intellectual limits of self and tap into alternate paths of enlightenment. "Historically, meditation has been practiced for at least three thousand years since the dawn of Indian yoga and is a central discipline at the contemplative core of each of the world's great religions."[17] Mystics, defined as ones who believe in the spiritual apprehension of truths that are beyond the intellect, are perhaps the best-known of these searchers for enlightenment.

---

[17] Shauna L. Shapiro, Roger Walsh, and Willoughby B. Britton, "An Analysis of Recent Meditation Research and Suggestions for Future Directions," *Journal for Meditation and Meditation Research*, Vol. 3, 2003, 69.

The objective of mystics is *transcending self* to achieve unity with all of creation, to see all things as one, to experience a reality that transcends anything that the mind's intellect could conceive. This journey typically begins with transcending concerns about worldly affairs, then words, ideas, and images, and finally self—a step-by-step nonattachment with the everyday world of space and time.[18]

What happens when we try to look at our world beyond self? Current neurological findings about mystical experiences are especially revealing in this regard. Andrew Newberg, a radiologist at the University of Pennsylvania, and fellow researchers, using a brain imaging technique called SPECT (single-photon emission computed tomography), studied a group of Tibetan Buddhist monks as they meditated for approximately one hour. At the moment they reached a transcendental high, they were asked to pull a kite string to their right, releasing an injection into their veins of a radioactive tracer. By injecting a tiny amount of radioactive marker into the bloodstream of a deep meditator, the scientists were able to see how the dye moved to active parts of the brain. Later, once the subjects had finished meditating, the regions were imaged and the meditation state compared with the normal waking state. The scans provided remarkable clues about what goes on in the brain during meditation. There was an increase in activity in the front part of the brain, the area that is activated when anyone focuses attention on a particular task. In addition, there was a notable decrease in activity in the back part of the brain, or parietal lobe, recognized

---

[18] Karen Armstrong, *A History of God: The 4000-Year Quest of Judaism, Christianity and Islam* (New York, NY: Ballantine Books, 1993), 220

as the area responsible for orientation. This finding reinforced the general suggestion that meditation leads to a lack of spatial awareness. During meditation, people have a loss of the sense of self and frequently experience a sense of no space and time, and that was exactly what Newberg and his fellow researchers saw.[19] (Later research showed that the complex interaction between different areas of the brain that occurred while the Buddhist monks were meditating resembles the pattern of activity that occurs during the spiritual practices of other faiths.)

An interpretation of what occurs during such intense spiritual experiences, from the dual-reality perspective presented above, is that the monks studied at some point no longer received the projected images (including the projected image called "self") of our everyday world, the world of elementary particles, and became aware, in some sense, only of a world beyond, usually referred to by mystics as the "All."

This interpretation is supported by an observation by quantum physicists about the nature of the quantum world of elementary particles. As shown by the so-called two-slit experiment described above, matter presents itself to us as *both* a particle and a wave—all particles have both particle and wave characteristics—*but we must make a choice as to whether to "observe" matter at any given time as a particle or a wave.* (The term "observe", which I use for want of a better term, is, in its usual sense, misleading since it implies that something was there to be observed before we looked for it; as you now know, experiments show that the object observed *is brought into being*

---

[19] Newberg, D'Aquili, and Rause, 1-6.

by our attempt to observe it.) That choice is made by awareness; if we attempt to "measure"—to become aware of—an elementary particle, we collapse the particle wave function and measure the presence of a particle. In the same way, our everyday world of elementary particles exists only if we are aware of it. The monks, in the final stages of meditation, were required to make a choice to experience either the projected images of our everyday world, a world "composed of" elementary particles,[20] or to move beyond that world to the All and they chose the All.

What can mystics tell us that might convince the doubters among us that they have indeed visited another world—what do they learn when they look beyond our everyday world? Mystics invariably assert that they find an all-encompassing truth at the moment of peak meditative experience when unity with the All occurs—but they have difficulty telling others what they have learned.[21] This "truth" that the mystics discover does appear to be *information*—it "informs" those who experience it since research shows that they are changed in rather profound ways by it. Shapiro et al. cite research studies which show that meditation is associated

---

[20] Some might object to the use of the phrase "composed of elementary particles" since this language seems to imply an overarching physical object, and would prefer language that implies an overarching information storage system containing information-theory units such as bits or bytes. In fact, however, in our everyday world we are presented with (virtual) physical objects and must deal with them as such.

[21] Evelyn Underhill, *Mysticism: The Nature and Development of Spiritual Consciousness* (Oxford, Eng: Oneworld Pubs., 1993), 4.

with significant psychological, physiological, and therapeutic effects. These effects include positive changes in specific mental qualities such as concentration and calm, and in emotions such as joy, love, and compassion, a clearer understanding of oneself and one's relationship to the world, and a deeper and more accurate knowledge of consciousness and reality.[22] If what the mystics learn has such a powerful effect on them, why are they unable to tell us what they have learned?

Part of the answer may be in the nature of the meditative experience. Mystics, to achieve a peak meditative state, must transcend words, ideas, images, and, finally, self. It is unlikely, therefore that what they learn would be framed in terms of these conceptual devices or that they could use these conceptual devices to tell others what they learned during their experience.

Another part of the answer may be our own limited capacities for reasoning. If we take the mystics at their word, to experience the "All" is to know everything and, in our everyday world, to know everything is to know nothing.[23] Within our limited capacities to reason, to avoid being overwhelmed with information, we must group our bits of information into a myriad of classes or categories—for example, we assemble animals into groups of similar animals, give names to animal types, and use these types in our thinking. Classification is an unavoidable requirement for

---

[22] Shapiro, op. cit., 86,

[23] John Barrow, in *Theories of Everything: The Quest for Ultimate Explanation* (Oxford, Eng.: Oxford University Press, 1991), 210, said: "No Theory of Everything can ever provide total insight. For, to see through everything would leave us seeing nothing at all."

thinking and learning. But when we group these bits of information and use these groupings in our analyses, we lose vast amounts of information about the unique properties of individual items in the group, in the example, about the animals themselves. This loss of information means that an all-encompassing truth is not within reach of our reasoning capacities and thus mystics cannot convey the information they learn to our scientists or to us. (Note that this loss of information likely is the everyday-world manifestation of the information loss that occurs at the quantum level with collapse of the wave function, a point to be discussed later in this chapter.)

The experiences of mystics suggests that the all-encompassing truth that they encounter during the mystical experience far exceeds our ability to use information. However, mystics are changed by the experience, as indicated above, and must, therefore bring to their everyday lives new information of some kind. What form might this new information take? Perhaps what the mystics learn is, for them in their everyday lives, both syntactic and connotative in nature. It is syntactic in that it serves as a guide to assembling bits of information but in itself conveys no semantic information. And it also may be connotative in nature—it has to do with, not literal meaning, but the feelings things arouse when we think of them.[24] Just as a painting conveys truths not expressible in words, so too may the mystical experience convey truths not expressible in words. Memories

---

[24] Connotative meaning also is discussed in connection with the semantic differential in Chapter 5.

of the mystical experience, however, may serve as a guide to a search for truth and allow the mystics to more nearly approach truth in their everyday world than would have been possible without the experience.[25]

Let's now shift our attention to that everyday world.

Knowing Our Everyday Virtual World

I have argued above that our everyday world is intangible or virtual in nature. This fact has important implications for knowing that world.

Implications of a Virtual World

Three important implications follow on from acceptance of the proposition stated above, that our everyday world is an intangible, virtual world of interconnected thoughts and feelings constituting a script for the drama of our lives. These implications of a virtual world are what I have chosen to call the "pointillist principle," the "information-world postulate," and the "logical imperative." They are needed for understanding our everyday world.

The Pointillist Principle

Upon acceptance of the intangible-world proposition, it becomes obvious that elementary particles, which have no tangible

---

[25] A similar experience in our everyday world is perhaps the calming experience of a walk in a natural setting with a predominance of fractal designs. Interestingly, Underhill (p. 191) indicates that an affinity for natural settings is an obvious feature in the history of the mystics.

existence,[26] are *no longer strange objects quite unlike our everyday world of physical objects*; rather, *they are the expected and obvious building blocks of a virtual world.* Elementary particles, that is to say, *share the same intangible existence as do the macro objects of our everyday world.* It follows that if elementary particles are the building blocks of our everyday world, *the same rules apply to the macro world as to the quantum world; therefore, findings made through observations at the quantum level can be applied at the macro level.*[27] I will refer to this inferential process as the "pointillist principle" because of its similarity to the process by which we draw conclusions from observations of the dots in a pointillist or divisionist painting that may be applied to the painting itself, and vice versa. (Pointillism uses dots of paint applied in patterns. It relies on the ability of the eye and the mind of the viewer to blend the dots to form an image. Its variant form, divisionism, uses a pure color in each dot, requiring the viewer to combine optically the different pure colors to obtain a fuller range of tones.)[28]

That characteristics of our universe as a whole can be seen in its parts also appears to be required by the holographic nature of our universe. If, as many scientists now believe, the universe

---

[26] Note, however, that this is not equivalent to saying that the whole is simply the sum of its parts. This point will be discussed in Chapter 2 in the context of a cosmic blueprint.

[27] "Although quantum effects have been observed only in the microworld of atoms and their constituents, in principle quantum physics should apply to everything." Davies, *The Cosmic Blueprint*, 5

[28] See "Pointillism," Wikipedia, November 23, 2011, and "Divisionism," Wikipedia, October 11, 2011.

is holographic,[29] it is subject to the holographic principle, which states that the description of *any* given volume of space in the universe can be thought of as encoded on a boundary to the region.[30] It further appears that a one-to-one correspondence may exist between information contained on the boundary, expressed in terms of quantum theory, and information contained in the interior of the space, expressed in terms of a quantum gravity theory (perhaps string theory).[31] This one-to-one correspondence of information would seem to require, throughout the universe, that characteristics of the whole may be seen in its parts. Thus holographic theory supports the pointillist principle, that is, that all findings regarding the characteristics of particles, building blocks of our intangible world, appear in some form at the macro level.

The pointillist principle leaves to us to discover how these quantum-level findings are manifested in the macro world. I will apply a number of these quantum-level findings to our macro-level everyday world in later sections of this chapter as well as in later chapters of the book.

---

[29] See "Leonard Susskind, Bad Boy of Physics," interview by Peter Byrne, *Scientific American*, July 2011, 80-83; Jacob D. Beckenstein, "Information in the Holographic Universe," *Scientific American*, July 14, 2003; and Juan Maldacena, "The Illusion of Gravity," *Scientific American*, November 2005, 56-63.

[30] Peter Byrne, interview with Leonard Susskind, *Scientific American*, July 2011, 80-83, 80.

[31] Maldacena, 61.

The Information-World Postulate

If our everyday world is an intangible world of interconnected thoughts and feelings brought into existence by our awareness, then the world consists of, is made of, information. That thought was expressed by two eminent scholars.

René Descartes, considered by many to be the father of rational thought and modern philosophy, felt that our reality is confirmed by our thinking. Writing in 1637, he said:

> I am, I exist—that is certain; but for how long do I exist? For as long as I think; for it might perhaps happen, if I totally ceased thinking, that I would at the same time completely cease to be. …. I am, therefore, to speak precisely, only a thinking being, that is to say, a mind, an understanding, or a reasoning being ….[32]

And John Wheeler, one of the most widely-respected physicists of the Twentieth century, who suspected that reality exists not because of physical particles but rather because of the act of observing the universe, said in 2002:

> Information may not be just what we learn about the world. It may be what makes the world.[33]

By our thoughts we bring our world into existence; thus the basic material of our intangible everyday world is *information, not matter*. This has profound consequences for many of the basic ideas

---

[32] René Descartes, *Discourse on Method and Meditations*, Translated with an Introduction by Laurence J. Lafleur (New York, NY: Macmillan Publishing Co., 1960) (Originally published in 1637).

[33] Folger, interview with John Wheeler.

of physics, a point we will address in various ways throughout the book. The most direct implication may be for entropy (the degree of disorder or randomness in our multiverse)—see Question 4 at the end of this chapter.

The Logical Imperative

A final implication of acceptance of the intangible-world proposition—in my opinion equally important—is that acceptance of this proposition may help to explain how the knowledge of our everyday world is structured. Our everyday world is believed to be governed over its entire extent by universal laws of physics.[34] In this everyday world, the scientific method, using the language of mathematics and the process of inductive reasoning—that is, the process of deriving general principles or premises from our observations of particular facts or happenings—is the primary tool for accumulating knowledge. And this tool has been strikingly successful in helping us to understand our universe. In an article widely accepted (and quoted) by both mathematicians and physicists, Eugene Wigner has called the appropriateness of the language of mathematics for the formulation of the laws of physics

---

[34] See P.C.D. Davies, "Where Do the Laws of Physics Come from?" scribd. com/doc/.../Where-Do-the-Laws-of-Physics-Come-From, October 8, 2008.

"a miraculous gift ... which we neither understand nor deserve."[35] The scientific method is so remarkably successful at explaining our everyday world that it seems very likely that the *organizing principle* used by the mind to assemble and present to us our everyday world is the very rules of logic and mathematics we use in inductive reasoning. Because the organizing principle used by the mind to assemble and to present to us the script for our everyday lives appears to be the logic of inductive reasoning, I have called this characteristic of our everyday world the "logical imperative": our everyday world is *inherently logical* in its structure—*it cannot be otherwise*. It goes without saying that the scientific method should be successful at explaining our world if the logic we use to understand that world is the logic used to present that world to us.[36]

Let's look now at that everyday world.

What Is Our Everyday World?

Until quite recently, knowing our everyday world would have meant knowing the world we see every day, what we know as our universe. The many-worlds interpretation of reality calls into

---

[35] Eugene P. Wigner, "The Unreasonable Effectiveness of Mathematics in the Natural Sciences," in *The World Treasury of Physics, Astronomy, and Mathematics*, Timothy Ferris (ed.) (Boston, MA: Little, Brown, 1991), 526-540, esp.540. For a more complete discussion of this point, see Mario Livio, "Why Math Works," *Scientific American*, 305, no. 2, August 2011, 80-83.

[36] This may help to explain why we enjoy natural settings (which contain fractal-geometric designs), music, or great paintings (which typically obey implicit mathematical rules) and other mathematically-structured designs.

question this assumption. This interpretation *has changed our view of the contents of time and space in profound ways and has raised fundamental new questions about the limits of human knowledge.*

The Many-Worlds Interpretation of Reality

A remarkable theory developed by Hugh Everett III, called the many-worlds interpretation (also called the many-universes interpretation) of quantum mechanics, suggests that the vast majority of our everyday world goes unseen by us but nevertheless is a quite real part of our world. This interpretation (MWI) is based on a "theory of the universal wave function," so called by Everett since *all of physics (thus at both the quantum and the macro levels) is presumed to flow from this function alone.*[37] Wojciech H. Zurek, a fellow at Los Alamos National Laboratory, explains the significance of Everett's universal wave function theory as follows: *"Everett's accomplishment was to insist that quantum theory should be universal, that there should not be a division of the universe into something which is a priori classical and something which is a priori quantum*[38]*."* (Note that this theory supports the pointillist principle, which states that elementary particles are the building blocks of our intangible world and, therefore, that all findings regarding the characteristics of particles *must* appear *in some form* at the macro level.)

---

[37] Hugh Everett III, *The Many-Worlds Interpretation of Quantum Mechanics: The Theory Of The Universal Wavefunction*, doctoral dissertation, Princeton University, 1957, 9. Everett's doctoral dissertation may be downloaded as a pdf file at www.pbs.org.

[38] Peter Byrne, "The Many Worlds of Hugh Everett," *Scientific American*, December 2007, 98-105, 104

The many-worlds interpretation views reality as a many-branched tree, wherein every logically possible quantum outcome is realized. (Recall that if the logical imperative is correct, it does not allow histories for us that are not logically possible, given who we are and our past histories.) Prior to the many-worlds interpretation, reality had always been viewed as a single unfolding history. In the many-worlds interpretation, all logically possible alternative histories and futures are considered to be real, each occurring in an actual "world" (or "universe"), each with its own observer.[39] In his doctoral thesis, Everett wrote: "all elements of this superposition are equally 'real.' Only the observer state has changed.[40]" By superposition, Everett means all theoretically possible worlds.

In lay terms, many-worlds posits that as we make certain critical choices in life, branching occurs such that a number of worlds emerge, one for each of the choices we could have made. Everything that could possibly have happened in *our* future, but will not, will occur in the future of some other world and everything that could possibly have happened in *our* past, but did not, has occurred in the past of some other world.

With the coming of the many-worlds interpretation of quantum mechanics, our world of time and space has become much more complex. We live not only in a universe but also in a "multiverse" in which our universe is only one of many universes. Knowing this,

---

[39] See Byrne, 101-102.

[40] Everett, 116-117. Particles are said to be in superposition when they exist in all their theoretically possible states as determined by the wave functions of the particles.

what opportunities are now available to us that were not available when we understood our universe as all that existed? For example, is there information in the alternative histories of those other worlds that we might be able to retrieve for use in our own world? Several quantum-level phenomena or experiments suggest that the answer is "yes" to the first part of the question—useful information is there.

Do Other Worlds Contain Information Useful to Us?

We will look at two experiments, in particular, that indicate information is available to us in the multiverse outside our own world.[41] Then we will take up the question of whether and how we might retrieve this information in a form suitable for application in our own world.

Wave Function Collapse and MWI

We know from the results of modified versions of the two-slit experiment discussed above that interaction by a wave with an observer causes what is called "collapse of the wave function" and the measurement ("observance") of an elementary particle. Wave function collapse is the phenomenon in which a wave function which initially satisfies a wave equation expressing various possible

---

[41] The so-called Feynman mirror experiment also indicates that information in the alternative histories of those other worlds might be retrievable for use in our own world. We will touch on this experiment later in the book. See Richard P. Feynman, *QED: The Strange Theory of Light and Matter* (Princeton, NJ: Princeton University Press, 1985), 49- 54 and 84-85.

states for an elementary particle—the particle as described by the wave equation is said to be in "superposition" of many different possible states—appears to reduce to a single one of those states after interaction with an observer. In doing so, it yields an outcome (we detect an elementary particle) and when this happens, no measurement of the other states can subsequently be made. From our standpoint, a loss of information has occurred.[42] Applying the many-worlds interpretation, this information is not lost but simply appears in a different form elsewhere in the multiverse where the particle appears in all its possible states.

In our daily lives, when we make a choice that causes branching of the many-worlds reality tree, wave function collapse brings into being a new world for us. We bring with us to our world *only* the information relevant to all of the past choices that brought us to our present quantum state together with the information relevant to all the future choices now logically possible given this latest choice. From our perspective, information was lost when we made our choice and started a new branch of the reality tree. However, there is no information loss in the "universal wave function," that is, in all the many worlds of our counterparts in those other worlds who made different choices, taken together. The information associated with other possible choices is available to our counterparts in other worlds but not immediately available to us. There is, therefore, information out there in those many worlds that is potentially useful to us—for example, what would we have experienced and learned on these "roads not taken."

---

[42] See Zukav, 99, and George Johnson, *Fire in the Mind: Science, Faith, and the Search for Order* (New York, NY: Alfred A. Knopf, 1995), 159.

And, parenthetically, we now see why science, as currently practiced, cannot have a "theory of everything". The information contained in our world (one of the many) is far too limited for such a theory. The "everything" we thought we were about to grasp as we planned for the theory of everything was not even close to being the "everything" it implies. For a theory of everything we need to access the vast stores of information "out there" beyond our everyday world.

The Delayed-Choice Experiment

A second experiment that indicates that information is available to us in the multiverse outside our own world is the delayed-choice experiment of John Wheeler. This experiment is a variation on the two-slit experiment. You will recall that in the version of the two-slit experiment conducted with individual particles, photons (particles of light) passing through two slits in a thin plate exhibit wave-like properties. However, placing detectors at the slits to determine which slit a particle is passing through achieves a startling result—it destroys the wave-like interference pattern on the screen behind the slits, indicating that the photons were now behaving like particles. The behavior of the photons thus was changed depending on whether or not they were observed.

The delayed-choice experiment was first proposed by John Wheeler in 1978.[43] In Wheeler's version of the two-slit experiment,

---

[43] John Archibald Wheeler, "The 'Past' and the 'Delayed-Choice' Double-Slit Experiment," in *Mathematical Foundations of Quantum Theory*, A. R. Marlow (ed.) (New York, NY: Academic Press, 1978), 9-48. This marvelously insightful piece by Wheeler is well-worth reading.

the method of detection used in the experiment (from our perspective) can be changed after a photon passes the two slits, so as to delay the decision of whether or not to detect the path of the particle until that path was already taken. Wheeler's hunch when he proposed this experiment, a hunch later confirmed by experiment, was that the universe is built like an enormous feedback loop, a loop in which we, by observing the universe, contribute to the *ongoing creation* of *not just the present and the future but the past as well.*[44] Wheeler described the question his proposed experiment was to explore as follows:

> Partway down the optic axis of the traditional double-slit experiment stands the central element, the doubly-slit screen. Can one choose whether the photon (or electron) shall have come through both of the slits, or only one of them, after it has already transversed this screen? That is the new question raised and analyzed here.[45]

As Wheeler himself expressed what confirmation of his hunch would mean:

> *No phenomenon is a phenomenon until it is an observed phenomenon.* In other words, it is not a paradox that we choose what *shall* have happened after *it has already happened.* It has not really happened, it is not a phenomenon, until it is an observed phenomenon. (Emphasis Wheeler)[46]

---

[44] Folger, interview with John Wheeler. We create (in the present time) the past? This statement has not received the attention it deserves.

[45] Wheeler, 9.

[46] Wheeler, 14.

When Wheeler's hunch was put to the test in experiments carried out between 1984 and 2007, the photons behaved as elementary particles or as waves, depending on whether detectors were in use or not, *before the experimenters had made up their minds about whether to use the detectors!*[47]

(Note that this experiment demonstrates once again that it is our attempt to observe that brings the elementary particles of our world into existence confirming the dual-reality presumption that the everyday world we experience is never "out there somewhere"—it is always and only in our minds!)

Applying the pointillist principle, the experiment would seem to support the view that we can search both the future and the past for information that we will not otherwise experience in daily life. As John Wheeler put it, we are tiny patches of the universe looking at itself[48]—and therefore we know what we did in both the future and the past. (Note that Wheeler is not reluctant to interpret these findings at the quantum level as having everyday-world consequences, in agreement with what I have called the "pointillist principle.")

---

[47] The first of these experiments were conducted by C. O. Alley, O. Jakubowicz, C. A. Steggerda and W. C. Wickes, "A Delayed Random Choice Quantum Mechanics Experiment with Light Quanta," in *Proceedings of the International Symposium on the Foundations of Quantum Mechanics, Tokyo, 1983*, S. Kamefuchi et al. (eds.) (Tokyo, Japan, Physical Society of Japan, 1984), 158; cited in Davies, *The Cosmic Blueprint*, 174.

[48] Tim Folger, interview with John Wheeler. See also Zukav, 87, for a description of our "self-actualizing universe."

Can We Retrieve that Information?

If the many-worlds interpretation as I have represented it above is correct, there would seem to be information in the multiverse that is potentially useful to us in our everyday lives. Can we retrieve useful information from the multiverse? Max Tegmark, a cosmologist who has written extensively about the many-worlds interpretation,[49] says that to insist that alternative histories are separated is a misrepresentation of the many-worlds interpretation and also is inconsistent with the Everett postulate.[50] And David Deutsch, one of the most prominent proponents of MWI, says that the phenomenon of interference (associated with the wave-like behavior of a particle) in the two-slit experiment suggests that the photons we measure (those that are found to be a part of our world) are not wholly partitioned from the photons in other worlds.[51]

That we focus on one choice only is no doubt necessary for our world to emerge and accounts for the failure of wave functions associated with other choices and other worlds to collapse; this circumstance therefore sets limits on our ability to interface with our counterpart worlds. If we are nevertheless able to search for information in those other worlds, how might this done?

Perhaps we *do* search for information in those worlds and are unaware that this is what we are doing. Newberg et al., whom we

---

[49] See space.mit.edu/home/tegmark/crazy.html.

[50] Max Tegmark, "The Interpretation of Quantum Mechanics: Many Worlds or Many Words?" September 15, 1997, 2, available online at sns.ias.edu/~max/everett.html

[51] David Deutsch, *The Fabric of Reality*, 45.

referenced above concerning their study of meditation by Tibetan monks, suggest that what happens during meditation leading to a peak meditative state (when *unity* with the "All" occurs) is a movement along a *unitary continuum*; they explain the movement in neurological terms as follows:

> The intensity of those unitary states depends upon the degree of neural blockage. Since the degree of that blockage can increase by any increment, and theoretically until there is a total blocking, a large spectrum of increasingly unitary states is possible. We call this span the unitary continuum. The arc of this continuum links the most profound experiences of the mystics with the smaller transcendent moments most of us experience every day, and shows that, in neurological terms, the two are different essentially by degree. [52]

Evelyn Underhill, writing without benefit of brain imaging studies (her book *Mysticism* was first published in 1911), but herself a mystic with extensive, research-based, knowledge of mysticism, reached the same conclusion as Newberg, D'Aquili, and Rause— that both mysticism and creative thinking are linked as a movement toward a unity and draw their meaning from outside the range of our daily experience.[53]

Transcendent experiences of daily life may or may not put us in contact with our counterpart selves in other worlds, but this seems

---

[52] Newberg, D'Aquili, and Rause, 115.

[53] Evelyn Underhill, *Mysticism: The Nature and Development of Spiritual Consciousness* (Oxford, Eng: Oneworld Pubs., 1993), 63.

to be a reasonable assumption if the many-worlds interpretation itself is correct. I will show in Chapter 3 that MWI should be accepted as correct and I will continue here with the assumption that our everyday transcendent experiences are putting us in contact with our counterpart selves in those other universes.

A transcendent experience of daily life, as I am using the term, is an experience that lies beyond the range of normal human experience and cannot be understood by ordinary reasoning but does not transcend self and therefore does not attain the peak meditative state of the mystic. These transcendental experiences appear to be a fairly common experience in our daily lives. A landmark study, carried out in 1975 by sociologist Andrew Greeley for the National Opinion Research Center, asked respondents: "Have you ever felt as though you were very close to a spiritual force that seemed to lift you out of yourself?" Remarkably, more than 35 percent of the study's population answered that they had.[54]

From the dual-reality perspective, the relationship between the intense spiritual experiences of mystics and those less intense transcendental experiences on the Newberg et al. continuum appears to be this: the *objective* of mystics is *transcending self* to achieve unity with an All, while the (spoken or unspoken) objective of most transcendent experiences of daily life is to solve some problem or achieve some objective of everyday life while *not transcending self*. Let's look at some examples of the latter kind of transcendental experiences.

---

[54] Cited in Newberg, D'Aquili, and Rause, 107. See also Joan L. Waldron, "The Life Impact of Transcendent Experiences with a Pronounced Quality of Noesis," *The Journal of Transpersonal Psychology*, 1998, Vol. 30, No. 2, 103-134, esp.103.

Transcendent Experiences of Everyday Life

As stated above, the spoken or unspoken objective of our everyday life transcendent experiences is to discover information relevant to some concern of everyday life. This daily-life concern appears to serve as *an anchor to the self and its extended world* (its multiverse), allowing the mind to enter a trancelike state without transcending self. As noted above, this type of creativity appears to characterize all of our lives to some extent.

Three instances of deliberate trance-enabled creative thinking have been studied sufficiently to allow us to conjecture about what is happening during the activity. These are the use of imagination by 1) creative thinkers who think imaginatively while in a self-induced trancelike state, 2) creative thinkers who think imaginatively while in a hypnotically induced trancelike state, and 3) creative performers who perform imaginatively while in a self-induced trancelike state.

Thinking Creatively while in a Self-induced Trancelike State

Some persons spend much or most of their time *thinking imaginatively.* These include both artists and scientists of various kinds: architects, designers, inventors, poets, and writers. Having worked most of my adult life among creative people, my experience has been that very creative people do two things very well. First, they have done the preparatory work necessary for the creative task: artists have developed the skills needed for their craft, scientists have carefully defined and fully researched the problem they are addressing. Second, they are able to focus intently on a problem or desired end result and to turn off other parts of the everyday world, as though in a trance, while the issue of concern continually drifts in and out of their minds. Experiments have shown the powerful effect of "zoning out" on creativity.[55]

From the dual-reality perspective, here is what seems to be happening when creative thinkers use their imaginations to aid their thought processes. Thinking in our everyday world typically is restricted to *rearranging* knowledge we already have, which can, in itself, be a creative process. In the many worlds of the universal wave function, however, where time doesn't pass but *is*, a search for answers is not so restricted. (To help understand this feature of time, note that in our four-dimensional world, we do not say that the three dimensions of space pass, they simply *are*; similarly, in the many worlds of the universal wave function, time does not pass, but *is*—there is no distinction between past and future.) Creative thinkers follow one or more logic trails into the future in a search for answers to the problem being researched.

Answers arrived at by such transcendental experiences appear not to be remembered information but are thoughts that have not occurred to us before, and such new thoughts typically feel like discovery, not like a created product or a newly remembered idea. The ancient Greeks saw creativity as discovery, not as our modern concept of "making something new." As one of the great thought experimenters, Plato, put it: people are put into the peculiar position

---

[55] See Carl Zimmer, "The Brain: Stop Paying Attention: Zoning Out Is a Crucial Mental State: Researchers Say a Wandering Mind May Be Important to Setting Goals, Making Discoveries, and Living a Balanced Life," | Mind & Brain | *Discover Magazine*, 06.15.2009

of discovering things that seemed to have been there all the time, unrecognized or forgotten, in thoughts buried most deeply in the most mysterious recesses of the mind.[56] Asked in *The Republic*, "Will we say, of a painter, that he makes something?" [Plato] answers, "Certainly not, he merely imitates."[57] And Einstein, one of the most imaginative and famous thought experimenters of modern times, stated (with coauthor Infeld) a like idea: "Physical concepts are free creations of the human mind, and are not, however it may seem, uniquely determined by the external world."[58] (Implicit in this statement is the inference that more than discovery is occurring in the thought experiment: perhaps the scientist also is *creating* the physical reality he or she imagines, an idea later echoed by John Wheeler.[59])

Roger Penrose describes this process for mathematicians in this way:

> When one 'sees' a mathematical truth, one's consciousness breaks through into this world of ideas, and makes direct contact with it . . . . When mathematicians communicate, this is made possible by each one having a direct route to truth, the consciousness of each being in a position to perceive mathematical truths directly, through this process of 'seeing.' [60]

---

[56] Cited in Cohen, ix.

[57] "History of the Concept of Creativity," *Wikipedia*, 12 May 2014.

[58] Einstein and Infeld, 31.

[59] Folger interview with John Wheeler.

[60] Roger Penrose, *The Emperor's New Mind: Concerning Computers, Minds, and the Laws of Physics* (Oxford University Press, Oxford, 1989), 428.

As we try to understand what is happening during creative thinking by artists and scientists, it is helpful to look at three aspects of the thought process. First, what is the role of the "thought experiment"? Second, do the answers found in the multiverse by creative thinkers seem to come from far in the future and from histories far afield from the current history of the thinker or from histories near at hand to the thinker's current history. Finally, do artists and scientists find and express the truths they discover in different ways?

Scientists and the thought experiment.

One type of creative thinking by scientists that deserves special attention is the "thought experiment." This form of experiment deserves special mention for it has been responsible for many of the **greatest** discoveries of all time. Martin Cohen, in his book *Wittgenstein's Beetle and Other Classic Thought Experiments*, describes a thought experiment as follows:

> The characteristic thing about both real and thought experiments is that you control and limit the circumstances and conditions for the test, so as to pick out just one variable or one unknown. The key difference is that in the latter, everything is set out not in reality but merely in the imagination. The circumstances are described, not created, and the action is imagined, not witnessed. Still, in a strange sort of way, the thought experimenter is just as much a witness (in a well-constructed thought experiment) as any laboratory scientist.[61]

---

[61] Martin Cohen, *Wittgenstein's Beetle and Other Classic Thought Experiments* (Malden, MA: Blackwell Publishing, 2005), ix.

Cohen lists Maxwell's Demon, Darwin's Theory of Evolution, Galileo's Gravitational Balls, Henri Poincaré's Alternative Geometries, Mach's Motionless Chain, Newton's Bucket, Socrates' Allegory of the Cave, and Einstein's Chasing the Light Wave as among the most famous of the thought experiments.

Thought experiments, which typically employ visual imagery as a mantra (a focal point or anchor for meditation) rather than a narrowly defined problem, may be especially appropriate for exploring the universe of the mind. In mystical experience at an intense level, images typically are transcended next to last (self is transcended last) in the mystic's step-by-step detachment from the everyday world. (See above.) At this step in the process, ideas no longer exist and rational thought processes have given way to more intuitive ways of understanding. A visual image, such as Einstein's thought experiment of racing a light beam leading to Special Relativity, may allow deeper penetration into that universe. A review of thought experiments, such as those described in Martin Cohen's book of classic thought experiments, suggests that imagery is associated with the more profound discoveries of science and the arts. And thought experimenters using imagery appear able to follow logic trails of discovery quite far into the future in a search for answers.

Answers from far and near.

If a search of our many-worlds alternative histories is indeed occurring, we would expect that some searches might extend far into the future or into histories far afield from the current history of the searcher. We would expect, then, that answers to these problems would not be logically extendable from currently available knowledge common to all or to most people, and therefore might be rejected or require considerable everyday-world thinking (rearranging knowledge) by many other scientists to see the logical connections between the answers and current knowledge. Richard Morris suggests that the scientific works of Galileo Galilei, Charles Darwin, Sir Isaac Newton, and Albert Einstein were so received. For example, many (not all) of Einstein's ideas were rejected initially by his peers because they were not logically extendable from then current knowledge. And in our own day, Morris suggests that the current works of Alan Guth and Stephen Hawking are examples among scientists of ideas that seem far ahead of their time.[62]

Staff writers for Online Universities suggest that artists, writers, and musicians as well are sometimes ahead of their time. They name the following masters of their craft whose work was not as beloved or highly regarded by people during their lifetime as it was by those in the generations that followed: Vincent van Gogh, Franz Kafka, El Greco, Johann Sebastian Bach, Henry David Thoreau, John Kennedy Toole, John Keats, Johannes Vermeer, Edgar Allan Poe, and Paul Gauguin.[63]

We would expect, also, to find that some of the works of artists and scientists from our distant past might appear quite prescient to our modern minds. The ideas of great scientists and philosophers

---

[62] Richard Morris, "Inventing the Universe," in *Creativity*, John Brockman (ed.) (New York, NY: Touchstone, 1993), 130-148.

[63] Staff Writers, "10 Incredible Artists Unappreciated in Their Time," *OnlineUniversities.com*, November 18, 2010.

of the past, such as Pythagoras (scientist best known for his works as a mathematician), Archimedes (mathematician), Aristotle (philosopher), Eratosthenes (scientist with important contributions in several fields) and Thales (philosopher), are still used today, over 2000 years after they developed their ideas.

Other answers of course might be found, not from far in the future or from histories far afield, but from the near future or from alternative histories directly related to current history. In fact, nearby alternative histories are likely contributing *directly* to our current history and are easily accessed by our imaginations. Again, borrowing an observation from the quantum world of elementary particles to apply to our macro world, while light is frequently said to travel in a straight line to an observer, Feynman showed with his well-known mirror experiment that while light waves travel *all possible paths* to an observer,[64] most of the waves cancel out (the trough of one wave encounters the peak of another wave) since their frequencies differ as a result of their following different paths to the observer. Waves traveling directly or *almost* directly (quantum effects likely enter here) to the observer have no other wave to interfere with them and so they actually arrive at the observer as a positive contribution. Light, then, doesn't really travel only in a straight line to the observer—it uses a small core of nearby space, and unless a mirror is large enough to accommodate this core group of waves, no light reaches the observer.[65]

---

64 John Gribbin, *Schrödinger's Kittens and the Search for Reality*, (Boston, MA: Little, Brown and Company, 1995), 92.

65 Feynman, 49- 54 and 84-85.

Employing the pointillist principle, which states that findings made through observations at the quantum level can be applied at the macro level, these findings from the mirror experiment suggest that most of us live, not in one narrow, closely bound, alternative history but in a somewhat indistinct cluster of closely related alternative histories that are easily accessed by our imaginations. (The pointillist principle likely is especially relevant here since light—electromagnetic force associated with the exchange of virtual photons—presumably is the medium used to transmit the holographic movies of our lives.) But just as we are unaware of the light reflected from the mirror outside of the core set of waves that actually reach us, in the same way we are only aware of those many-world waves of the universal wave function that collapse to bring our particular world into being.

Some, perhaps most, of the answers of creative thinkers are found in closely related alternative histories. These are perhaps those discoveries that produce the "aha!" moment as the immediately recognizable logical extension of the answer to current knowledge, the ones where we say to ourselves, "Why didn't I think of that before?"

Artists, scientists and truth.

A fundamental precept of the dual-realities theory is what I have called the logical imperative: the *organizing principle* used by the mind to assemble and present to us our everyday world is the very rules of logic and mathematics we use in inductive reasoning. If the scientific method is "built into" our world, is there a role for the arts in expressing truth. Or are the arts then illogical?

To me, and I'm sure to most of my readers as well, the poem that expresses to perfection and recalls without flaw our feelings is not illogical; rather, it is *meta-logical*—it captures a much more complete truth than does a mere logical description of the same ideas. It does this by drawing not only on factual information from our everyday world, but also on transcendental information retrieved by artists from outside our everyday, logical world.

In this regard, mathematicians perhaps bridge both worlds—the sciences and the arts. While their use of formulas requires adherence to the rules of a formal logic, mathematical elegance of their formulas frequently is used as a standard of truth by mathematicians.[66] For example, Shing-Tung Yau, a well-known geometer who was awarded the Fields Medal in 1982, says of models he has constructed of the universe's hidden dimensions: "No one can say for sure whether these models will ultimately prove correct. But the theory underlying these models, nevertheless, possesses a beauty that I find undeniable."[67] And Paul Dirac, the theoretical physicist whose aesthetic deliberations led him to construct a mathematically more elegant equation for the electron, stated: "It is more important to have beauty in one's equations than to have them fit experiment."[68]

---

[66] See Timothy Ferris, ed., *The World Treasury of Physics, Astronomy, and Mathematics* (Boston, MA: Little, Brown, 1991), p. 528.

[67] Shing-Tung Yau and Steve Nadis, *The Shape of Inner Space: String Theory and The Geometry of the Universe's Hidden Dimensions* (New York, NY: Basic Books, 2010), p. x.

[68] Davies, *The Mind of God*, p.176

Thinking Creatively while in a Hypnotically-Induced Trancelike State

Creative thinking while in a hypnotically-induced trance and creative thinking while in a self-induced trance, aside from the manner in which the trance is induced, appear to differ largely in the *depth of the trance*. The subject during hypnosis appears to be under a *much deeper* trance than is the typical creative thinker.

During hypnosis, the hypnotist directs the subject's thoughts, feelings, and behavior by instructing him or her to concentrate on particular images, ideas, or tasks. While under hypnosis, subjects are far more open than usual to suggestions and have the ability to focus intensely on a specific thought, feeling, sensation, or task. The brain state while under hypnosis appears to go beyond our usual attempts to imagine something. Research done at the University of Geneva suggests that hypnosis alters neural activity by rerouting some of the usual connections between brain regions; such neurological detours don't happen when subjects not under hypnosis imagine a scenario.[69] Preliminary findings suggest that hypnosis boosts activity in the brain's prefrontal cortex—a region responsible for various executive functions such as decision making and regulating attention—while suppressing activity in other brain regions, similar to the changes in brain activity that occur during meditation by mystics.

---

[69] Susan Gaidos, "The Mesmerized Mind," *Science News*, October 10, 2009, Vol. 176 Issue 8, 26-29

The most extensive series of experiments examining creativity during hypnosis was conducted by a physician, Dr. Linn F. Cooper.[70] In the usual course of Dr. Cooper's experiments, a simple trance state was effected by suggestions of sleep. Then Dr. Cooper would usually say, "Now, give me your attention, please. When I give you the starting signal, by saying 'Now', you're going to (you will) ..." followed by instructions about the experience the subject was to have. At the end of a predetermined time, Dr. Cooper would halt the trance by giving a prearranged termination signal, typically the word "blank," meaning the subject's mind was to become blank at that point.[71]

Dr. Cooper conducted a number of experiments in which subjects were asked to engage in creative mental activity while under hypnosis. One such experiment involved a subject who was employed as a secretary but had long been interested in dress designing and possessed considerable talent in the field. In her written account of her customary approach to designing a dress, she stated that when she has decided that she wants to design a dress, she begins a, usually, several-month process of looking at dresses in shop windows and pictured in publications, selecting and buying material, "playing" with the material and draping it on herself. Then she

---

[70] Linn F. Cooper and Milton H. Erickson, *Time Distortion in Hypnosis: An Experimental and Clinical Investigation* (Boca Raton, FL.: OTC Publishing Corp, 2004). (The book is divided into three parts; Dr. Cooper's research is reported in Part I.)

[71] For a detailed description of the experimental procedure used in the research, see Cooper and Erickson, 44-53.

generally starts cutting muslin patterns and when she has a basic pattern that pleases her, she cuts out the material and makes the dress, deciding on the details as she goes along. Sometimes she draws as she makes the dress and uses the drawing as "sort of a working tool" but never draws the entire design.

As a control, Dr. Cooper twice assigned to her in the waking state the task of designing a dress. In the first case, she left her desk at the end of a half hour and said, "I'm just wasting my time because I can't think of anything about anything. I don't have one single idea." In the second, she likewise produced nothing in the course of a half hour saying (in a written report), "I sat there and thought or, more precisely, tried to think, but the only things I thought of at all were newspaper or magazine pictures of dresses. I can't just sit down and design something, and I never attempt to because I know it is useless."

While in a trance state under hypnosis, she was asked in six separate sessions to design a dress. In each session, she was given the following instructions: "When I give you the starting signal by saying 'Now', you're going to design a dress. You'll have all the time you need." The subject was waked immediately after the completion of each trance activity and (with the exception of Task 1, where she was only asked to describe the design) was asked to draw a picture of the dress she had designed and to describe it briefly. These pictures and her descriptions of the design were quite detailed.[72]

---

[72] For drawings and descriptions of the dresses, see Cooper and Erickson, 139-143.

Asked to compare her performance in the waking and trance states she said:

> In the waking state I sat there for ages and tried every way I knew how to think of something and could not. While trying to think, I doodled bits of dresses I had seen pictures of, but I never did think of anything that ... seemed original in any way.
>
> In the trance state, my mind seemed clearer and I didn't have to try so hard. I had several ideas which I manipulated until I had what I considered the best combination in my mind. Later I recorded them on paper.

Other subjects were asked (examples are given in parentheses): to solve hypothetical problems (should a young girl, daughter of well-to-do parents, seek a job?); to prepare and deliver within the trance state a brief lecture (to a group of senior high school students on the career field of the subject); to make up a story (about "a horse, an old woman, and a ship"); or to produce a never-before-seen movie "short" (a comedy or animated cartoon). As in the dress-design case above, the reports were amazing in the fullness of detail, the careful and thorough consideration of the task, and the amount of reflection they indicated.

How do these subjects arrive at their creative ideas? Dr. Cooper says that, unless the hypnotist suggests otherwise, new ideas are created by building on remembered experiences, using skills and knowledge acquired prior to the hypnotic experience.[73] This, of course, could also be said of all creative experiences in our everyday

---

[73] Cooper and Erickson, 168.

lives. While the products created under hypnotist-induced trance were original with the subjects, they perhaps seem to the reader less spectacular and to lack the far-future focus often seen in creative thinking during self-induced trances. It should be noted, however, that when we wrote earlier of creative thinkers in self-induced trances we were thinking of accomplished professionals, of persons who "thought for a living" so to speak, whereas the hypnotic subjects studied by Dr. Cooper were not typically accomplished professionals but students at various levels of education. It is therefore not possible in this instance to make a meaningful comparison of the efficiency of these two methods for achieving creativity.

A closing thought: Dr. Cooper believes that the hypnotic trance experiences of subjects are as real as our everyday experiences.[74] If the reader accepts the dual-realities hypothesis presented in this chapter, that what we experience as everyday life is all in our minds, it is difficult to deny the reality of the hypnotic trance experience.

Performing Creatively while in a Self-Induced Trancelike State

When we use the word "creativity" we normally think of people who create and express ideas and feelings; we forget that people who accomplish physical tasks, especially those who are professionals, also are creative. These people include acrobats, actors, athletes, dancers, magicians, musicians, puppeteers, and singers.

Creative performers share the above-noted work habits of creative thinkers who use a self-induced trance to aid creativity: first, they spend many hours in learning about their profession and perfecting their skills, and second, while performing, they are able to

---

[74] See Cooper and Erickson, 18-19.

retreat into their own world, as though in a trance. The difference between creative thinkers and creative performers is that while creative thinkers are searching for new ways to synergistically arrange ideas and express feelings, creative performers must relate in new ways *their kinesthetic sense of their own bodies* to the space, objects, and other people involved in their performances.[75] This learning of new ways to perform is termed kinesthetic learning (or, alternatively, proprioception) defined as "the sense of the relative position of neighboring parts of the body and strength of effort being employed in movement."[76] The following description by Mary Wigman, considered one of the most important figures in the history of modern dance, of the way her dance the Pastorale, took form illustrates kinesthetic learning in dance.

> My Pastorale was developed in the following way: I came into my studio one day and sank down with a feeling of complete relaxation. Out of a sense of deepest peace and quietude I began slowly to move my arms and body. Calling to my assistants I said, "I do not know if anything will come of this feeling, but I should like a reed instrument that would play over and over again a simple little tune, not at all important, always the same one." Then with the monotonous sound of the little tune, with its gentle lyric suggestion, the whole dance took form. Afterwards we found that it was

[75] Michael Murphy and Rhea A. White, *The Psychic Side of Sports* (Reading, MA: Addison Wesley, 1978), 116.

[76] "Proprioception," Wikipedia, December 7, 2012.

built on six-eighths time, neither myself nor the musician being conscious of the rhythm until we came to the end.[77]

Creative performers then, like thought experimenters, must think using imagery and, crucially, the images *must include themselves.* These aspects of creative performance are extensively documented in the performance psychology literature, especially in its subfield sports psychology,[78] and are illustrated in the following descriptions by creative performers of the way they practiced and performed.

Billie Jean King, former professional tennis player, writes that she is at her best when she is playing totally by instinct. She says that on those super days,

I don't worry about hitting the ball and I hardly notice my opponent at all. It's like I'm out there by myself. I've talked with [other professional tennis players about this] and on their great days their attitude is exactly the same. I concentrate only on the ball in relationship to the face of my racket .... I appreciate what my opponent is doing, but in a

[77] From Mary Wigman, "Composition in Pure Movement," *Modern Music,* January-February, 1946, reproduced in Ghiselin, Brewster (ed.), *The Creative Process: A Symposium* (New York, NY: New American Library, 1954, 75-76.

[78] Murphy and White, op. cit. Excellent discussions of the research about the effects of mental imagery on athletic performance and the neural mechanisms that allow improved performance are Jeffrey J. Janssen and Anees A. Sheikh, "Enhancing Athletic Performance through Imagery: An Overview," 1-22, in Anees A. Sheikh and Errol R. Korn (eds.) and Imagery in Sports and Physical Performance: Imgery and Human Development Series (Amityville, NY: Baywood Publishing Company, 1994) Other useful materials are listed in the bibliography.

detached, abstract way, like an observer in the next room. I see her moving to her left or right, but it's almost as though there weren't any real opponent, as though I didn't know— and certainly didn't care—whom I was playing against.[79]

Jack Nicklaus, champion golfer, describes the way he visualizes a shot before he makes it:

I never hit a shot, even in practice, without having a very sharp, in-focus picture of it in my head. It's like a color movie. First I "see" the ball where I want it to finish, nice and white and sitting up high on the bright green grass. Then the scene quickly changes and I "see" the ball going there: its path, trajectory, and shape, even its behavior on landing. Then there's a sort of fade-out, and the next scene shows me making the kind of swing that will turn the previous images into reality. Only at the end of this short, private, Hollywood spectacular do I select a club and step up to the ball.

For the golfers among my readers, Nicklaus cautions that such visualization should never show a less-than-perfect shot. [80]

Sometimes performers see improvement in their performance from one mental exercise to the next. Lee Evans, a 400 meters Olympic champion and world record holder, used self-hypnosis and mental practice for several years in practicing for the 1968 Olympics. He visualized every step of the 400-meter race until he saw each stride he would take. By repeating this exercise again and

[79] Billie Jean King with Kim Chapin, *Billie Jean* (New York, NY: Harper and Row, 1974): 197.

[80] Jack Nicklaus with Ken Bowden, *Golf My Way* (New York, NY: Simon and Schuster, 1974) 79-80.

again, he says, his style and pacing got better and the overall flow of his performance was perfected.[81]

The creative ideas and images that promote kinesthetic learning of performers appear to be formed by imagining alternative future performances and selecting from among these performances those to be practiced imaginatively. Since discoveries made during creative efforts by performers are inherently constrained by the limits of the human body, we would not expect to find striking examples of sudden improvements in performance. But great performers repeatedly give such mental practice much credit for their success.[82]

Can the Dual-Realities Interpretation Be Proven?

Proof is an "is or is not" question that is established by evidence and has meaning *only* within our world of time and space where "things" (including ideas as unitary objects) exist. I believe it will always be impossible to prove definitively, within our everyday world, the dual-realities interpretation of reality, that is, that we live in a virtual world set within a, perhaps, more real, other world beyond time and space. If it is correct that the laws of physics apply everywhere in our multiverse—that they are coextensive with spacetime—then the virtual reality of our everyday world is complete; it leaves no loopholes that we might exploit to prove that we exist in tangible

---

[81] Murphy and White, 120.

[82] Murphy and White, 157.

form.[83] It is somewhat like trying to find out whether you are in the "Land of Liars" by asking those around you.

Indirect evidence that we live in an intangible multiverse world set in a larger world beyond time and space will be provided *if this hypothesis provides better ways to understand* some of the enigmatic ideas of science, such as the nature of time and time itself, wave function collapse, the space-traveling-twin saga, the many-worlds interpretation, Bell's theorem, particles in superposition, and what gravity is. I will discuss implications of the hypothesis for these topics and others in questions at the end of this chapter and in the remainder of the book.

And direct evidence to support (although not to prove definitively) the dual-realities concept may be provided if it can be shown that one of its key contentions is correct: the claim that information is available to us from beyond our everyday world. (Current applications of particles in superposition in leading-edge applications such as quantum computing and condensed matter physics using information from our counterpart worlds may make this question moot. This topic will be discussed in some detail in Chapter 3 below.)

We will discuss below current evidence in research studies to show that information is available in other worlds. Also in this section, we will consider an apparent paradox of the dual-realities interpretation that I will discuss under the heading: the "extra-body" problem.

---

[83] A possible exception within our world of time and space is the first millisecond of cosmic expansion and deep inside black holes, where the basic physics remains unknown. Martin Rees, *Just Six Numbers: The Deep Forces that Shape the Universe* (New York, NY: Basic Books, 2000), 158. However, in regard to black holes, see p. 63 of Juan Maldacena, "The Illusion of Gravity," *Scientific American*, November 2005, 56-63

Evidence that Information Is Available in Other Worlds

Evidence to support the claim that information is available to us from outside our own world might come either from quantum-level research or from macro-world research.

Quantum Level Research

At this writing, only one quantum-level experiment, John Wheeler's delayed-choice experiment, appears to be relevant to this question of whether information is available to us from beyond our own everyday world. You will recall from our discussion above of John Wheeler's version of the two-slit experiment,[84] that photons behaved either as elementary particles or as waves, depending on whether detectors were in use or not, *even when the experimenters had not yet made up their own minds about whether to use the detectors!* The experiment thus would seem to support the view that we can search both the future and the past for information that we will not otherwise experience in daily life. As John Wheeler put it, we are tiny patches of the universe looking at itself—and therefore we know what we did in both the future and the past.[85]

To the author's knowledge, no other experiments feasible with current technologies are being conducted to tell us whether information is available to us from beyond our everyday world. David Deutsch has proposed three thought experiments designed to

---

[84] Wheeler, "The 'Past' and the 'Delayed-Choice' Double-Slit Experiment," in Marlow, *Mathematical Foundations of Quantum Theory*, 9-48.

[85] Tim Folger, interview with John Wheeler. See also Zukav, 87, for a description of our "self-actualizing universe."

show that "interference" between many-worlds alternative histories occurs. One of these experiments should decide this issue but is not feasible with currently available technologies.[86]

Macro-World Research

Macro-world research also offers some support for the idea that we are able to search the future for information of relevance to us in a present time. All of the experiments cited below involve non-entranced persons in everyday life—persons not in a self-induced or hypnotic trance. (Note, however, that my interpretation is *not* the way the researchers interpreted the results of their experiments.)

In research by Amelia Hunt, subjects were asked to stare straight ahead with a ticking clock off to one side and to move their eyes over to the clock and make a note of the time when they had done so. On average, they reported seeing the clock about four hundredths of a second before their eyes actually arrived there.[87]

In research by A. Bechara et al., subjects began to choose advantageously *before* they realized which strategy worked best and

---

[86] David Deutsch, "Three Connections between Everett's Interpretation and Experiment," in *Quantum Concepts in Space and Time*, R. Penrose and C. J. Isham (Oxford, Eng.: Clarendon Press, 1986), 215-225

[87] Amelia Hunt, "An Interesting 100 Milliseconds in Which To Examine Vision and Attention," a seminar presentation at the Vision Sciences Lab, Harvard University, Wednesday, March 12, 2007, cited by Carl Zimmer, "How Your Brain Can Control Time," *Discover Magazine*, July 12, 2008.

also began to generate anticipatory skin conductance responses whenever they pondered a choice that *later* turned out to be risky.[88]

And a study by Chun Siong Soon et al.,[89] building on an earlier study by Benjamin Libet,[90] found that the outcome of a decision can be encoded in brain activity of the prefrontal and parietal cortex up to ten seconds before it enters awareness, a delay that the authors say presumably reflects the operation of a network of high-level control areas that begin to prepare an upcoming decision long before it enters awareness.

Studies using a design specifically intended to test the effectiveness of subjects in a self-induced trance or (better) a hypnotic trance at searching the future for information might indicate whether the assumptions of the dual-reality hypothesis are correct. The Bechara et al. research in particular would seem to lend itself to the use of entranced subjects while modifying the research design to ensure that the ambience was conducive to arriving at and maintaining a trance state. The research might look at how far in

---

[88] A. Bechara, H. Damasio, D. Tranel, and A. R. Damasio, "Deciding Advantageously before Knowing the Advantageous Strategy," *Science*, 275: 1293-95, 1997.

[89] Chun Siong Soon, Marcel Brass, Hans-Jochen Heinze, and John-Dylan Haynes, "Unconscious Determinants of Free Decisions in the Human Brain," *Nature Neuroscience* (Nature Publishing Group) 11 (5), April 13, 2008, 43–45.

[90] B. Libet, C. A. Gleason, E. W. Wright, and D. K. Pearl, "Time of Conscious Intention To Act in Relation to Onset of Cerebral Activity (readiness-potential): The Unconscious Initiation of a Freely Voluntary Action," *Brain*, 106, 1983, 623-642.

advance of conscious awareness of the best strategy persons in a trance state could choose advantageously as compared to non-entranced controls and what this would tell us about the mechanism causing this time difference.

Proof and the "Extra Body" Problem

It may appear to some readers that we already have reason to doubt the dual-reality hypothesis when used to interpret what happens during a hypnotic trance. If the hypothesis were true, then where are the products produced by hypnotic subjects during the time they are under a hypnotic trance? For example, the hypnotic subject who also was a dress designer says she made a drawing while in a trance preparing one of the designs; where is the drawing after the trance has ended? The explanation might be said to involve an "extra body" problem. It is well to remind the reader at this point that the dual-reality hypothesis involves the assumption that the worlds of subject and hypnotist are separate personal scripts emerging from their respective minds.

The hypnotist is in a world created by his awareness and that world includes the entranced subject's body in its entirety, that is, including his or her mind, lying *motionless* on a couch. Meanwhile, the subject is *also* in a different world, a world created by the subject's awareness, a world that does not include the hypnotist and his laboratory. And in that world the subject is producing a drawing. The body of the subject thus is in both of these worlds. Each of these worlds must obey its own logical imperative, and therefore no product produced by the subject during a hypnotic trance can suddenly appear in the hypnotist's laboratory. Upon waking, all that the subject can bring into the (now shared) world of hypnotist and subject is a memory of the hypnotic experience.

Beyond Evidence; How about Seeking Truth?

For many persons, an alternative to *proving* an assertion is to look for the fundamental *truth* of an assertion. If it is impossible to prove an assertion, an alternative is to search for a fundamental truth that shows the assertion to be true. That is, truth may be seen as an alternative to evidence-based proof.

As discussed in this chapter, many persons would contend that it is not possible to arrive finally at truth within our everyday world. Mystics, however, invariably assert that they find, *beyond* our everyday world, an all-encompassing truth at the moment of peak meditative experience when unity with the "All" occurs.[91] In a similar way, creative people who speak of "discovering" new ideas or new ways of expressing feelings typically do not require vindication of their discovery—they *know* the idea or art form expresses truth. Einstein is reported to have said, when asked what he would do if an experimental test of his general theory of relativity didn't agree with the theory: "So much the worse for the experiment. The theory is right!"[92] And mathematicians feel a sense of certainty or truth when they "see" a mathematical truth; Paul Dirac, the theoretical physicist who constructed a mathematically more elegant equation for the electron, which led to the successful prediction of the existence of antimatter (the positron was discovered two years

---

[91] Underhill, Underhill,, 4.

[92] Davies, *The Mind of God,* 175-176.

after Dirac constructed his formula), echoed these sentiments: "It is more important to have beauty in one's equations than to have them fit experiment."[93]

Mystics and creative people who discover their ideas in the universe of the mind have in common that they acquire their certainty through *experience rather than experiment*—they *experience* and they know based on that experience. Skeptics may scoff at this. They may prefer to stay within the confines of the incomplete truth of the observable portion of our everyday world to the, perhaps, unprovable truth of mystics and thought experimentalists. (Here I am referring to discoveries not subject to verification by laboratory experiment). Evelyn Underhill, herself a mystic, challenged this reluctance of skeptics to accept the testimony of those who have "seen" outside our universe. Writing in 1911, she says:

> [The mystics] should claim from us the same attention that we give to other explorers of countries in which we are not competent to adventure ourselves; for the mystics are the pioneers of the spiritual world, and we have no right to deny validity to their discoveries, merely because we lack the opportunity or the courage necessary to ... prosecute such explorations ....[94]

This argument is especially persuasive since people who experience genuine mystical states are credible persons who enjoy higher

---

[93] Davies, The *Mind of God*, 176.

[94] Underhill, Underhill, 4.

levels of psychological health than the public at large.[95] To those who choose not to believe the mystic, the gauntlet is cast: go to the "foreign land" and see for yourself!

As it becomes more and more difficult to verify our hypotheses experimentally, we may begin looking more and more to the universe of the mind for truths discovered by thought experimenters and mystics and this truth may obviate the need for proof.

Does it Matter?: Implications of the Dual-Realities Hypothesis

Whether the dual-realities hypothesis is true or not and whether people in general believe that they live in a virtual world will not make a difference in the lives of most of us. Just as the finding that the earth was not flat did not affect significantly the way people lived their daily lives, so too with the dual realities hypothesis. We are virtual beings living in a virtual world, a circumstance we cannot change. We will continue life much as usual.

For the scientific community however, it is always an important step forward to acquire new information about how our world works. If the dual-realities hypothesis is accepted, its impact on science, as indicated above, would be immense, allowing a fresh look at all of the major topics of science, such as the nature of time itself (to be covered in the next chapter), an interpretation of Bell's theorem, what gravity is, the nature of entropy, and the overlap of religious beliefs and science. In one way or another—either as part

---

[95] See Newberg, D'Aquili, and Rause, 108-109; see also A. Greeley, "Mysticism Goes Mainstream," *American Health*, Vol. 6 (1), 1987, 47-49.

or all of a chapter or as a response to a question—I will examine the implications of dual realities for each of these topics.

The most important implication, however, may be for the future of science itself and its role in the advancement of knowledge. Knowing that we live in a world of two realities and that logic works in only one of these realities, we may need to move beyond rational explanation. Commenting on the limits of science and a possible future path for the advancement of knowledge, Paul Davies says:

> We are barred from ultimate knowledge, from ultimate explanation, by the very rules of reasoning that prompt us to seek such an explanation in the first place. If we wish to progress beyond, we have to embrace a different concept of "understanding" from that of rational explanation. Possibly the mystical path is a way to such an understanding. I have never had a mystical experience myself, but I keep an open mind about the value of such experiences. Maybe they provide the only route beyond the limits to which science and philosophy can take us, the only possible path to the Ultimate.[96]

Other of the world's finest thinkers, including notable scientists such as Einstein, Pauli, Schrödinger, Heisenberg, Eddington, and Jeans, have also espoused mysticism.[97]

We perhaps need a new profession beyond those of physics and philosophy that pursues this higher level of understanding.

---

[96] Paul Davies, *The Mind of God*, 231-232.

[97] Davies, *The Mind of God*, 226.

This profession might be epistemology, the pursuit of knowledge of our world, perhaps anchored by Bertrand Russell's two ways of knowing—"knowledge by description" and "knowledge by acquaintance."[98] Its practitioners would be skilled at the practice of meditation and would give emphasis to thought experiments as a way to explore the frontiers of knowledge, *while not abandoning research skills and physical experiments.* Hopefully, in the teaching academies, just as the aim of an educator in our universities now is to have all students become adept at the scientific method, students would also do course work on meditation as well as practice meditation as part of their course work. Speaking of a new physics prompted by quantum mechanics, Zukav says, "Do not be surprised if physics curricula of the twenty-first century include classes in meditation."[99]

Students would be encouraged to do thought experiments as course assignments and would become comfortable with both acquiring knowledge and using their unfettered imaginations. This new profession may allow us to see that we have been living in a world of self-imposed ignorance similar to the other-imposed ignorance of Socrates' Cave Dwellers.

Socrates describes an imaginary group of people who have lived all of their lives chained to the wall of a cave, facing a blank wall. The people watch shadows projected on the wall by things

---

[98] See Bertrand Russell, "Knowledge by Acquaintance and Knowledge by Description," in *The Basic Writings of Bertrand Russell, 1903-1959*, Robert E. Egner and Lester E. Denonn (eds.) (New York, NY: Routledge, 1961), 217-224.

[99] Zukav, *The Dancing Wu Li Masters*, 327.

passing in front of a fire behind them and begin to ascribe forms to these shadows. The shadows are as close as the prisoners get to viewing reality. Once freed, the prisoners come to understand that the shadows on the wall do not make up reality at all, as they now can perceive the true form of reality rather than the mere shadows they saw as prisoners. Can we, like the Cave Dwellers, secure freedom from our own chains?[100]

In earlier times, the human race traveled to distant lands and found these lands strange but also found that crossing frontiers had many advantages and eventually seemed not strange but exciting. These pioneers expanded our world in ways that could not have been anticipated in advance. This I believe will be our experience as we learn to transition effortlessly between our dual realities.

Chapter Summary

Truth has meaning only in the context of reality. What is real? We usually assume we know what is real—our everyday world of apparently tangible objects that we can detect with our senses. Modern science suggests that we live in a world far, far more complex than this.

Quantum physics and brain imaging studies show that what we experience as reality may be a rendition of reality that is only in our minds. And a fairly recent theory tells us that this intangible world is itself only one of many alternative worlds, a quite real but largely

---

100 This description is from "Allegory of the Cave," Wikipedia, December 3, 2012. For a discussion of this allegory as a thought experiment, see "F is for the Forms Lost Forever to the Prisoners of the Cave" in Cohen, 30-32.

unseen multiverse. Further, since the images we experience as our everyday world are unlikely to originate in this intangible world itself, there must be a "world beyond" from which the script of our daily lives originates.

While our science has been remarkably successful in coming to know our own world—our universe—our knowledge of the multiverse and the world beyond is quite limited. Future advances in knowledge, and especially answers to "ultimate" questions, will largely depend on information we obtain from these worlds outside our own universe. The chapter explores various ways we can obtain this information as well as changes we might make in our fields of study to allow more ready access to this information in the future.

Questions

Question 1. Are the discoveries of ideas by creative thinkers, discussed in the chapter above, found in the subconscious minds of the thinkers?

An article by David Eagleman in Discover magazine attributes the discoveries of scientific ideas to thoughts emerging from an implicit memory stored in our subconscious minds. The article gives the example of nineteenth-century mathematician James Clerk Maxwell, who developed fundamental equations unifying electricity and magnetism. Maxwell, in a deathbed statement, declared that "something within him" had made the discoveries; that he actually had no idea how he'd achieved his great insights. Eagleman contends that discoveries of new ideas by creative thinkers such as Maxwell emerge as a memory from the subconscious mind.[101] Is the subconscious a more reasonable source for discovered ideas than the many worlds of the universal wave function?

As Eagleman contends, it has been well established by scientists that only a tiny fraction of the brain is dedicated to conscious behavior. The rest works behind the scenes and includes an implicit memory of our experiences, most of which cannot be remembered by the conscious mind. This inaccessibility of these memories to the conscious mind, according to the article, makes thoughts based upon these memories and emerging from this area of the mind seem like discoveries to us.

A problem with this explanation of why our creative thoughts seem like discoveries is this: If discoveries of new theories or concepts arise out of our memories, the discoveries would seem unlikely to have a future focus. Yet, this future focus seems almost a defining feature of the great advances in knowledge in the past—they *were* advances in knowledge. They were far ahead of the thinking at the time when they were first published and continued to be useful many years after becoming part of then-current given wisdom. For example, Einstein's relativity theories were well ahead of his time when first published in the early years of the 20th century and continue to be relevant today, nearly 100 years after their development.

If we wish to search the *future* for creative ideas we may do so in the many worlds of Hugh Everett where time does not pass but

---

[101] Eagleman, David. "The Secret Life of the Mind," *Discover*, September 2011, 50-53. This article is based upon Eagleman's book *Incognito* (New York, NY: Pantheon Books, 2011).

is and where our minds can freely roam time's landscape searching for creative ideas. To realize that we discover, that we do not create great ideas, is an humbling experience that keeps our own contributions in perspective—we search for truth, we do not create truth.

Question 2. If our world, including our bodies, is virtual, consisting only of information, might our bodies then be, essentially, computer programs? And if our bodies are computer programs, what are the implications of this for the advance of knowledge—for example, what might this mean for the field of health care?

I believe it is probably so that our bodies are computer programs. And if they are, perhaps disease and deformities frequently are flaws in our bodies' computer programs. In the future we may learn to identify and remove "programming errors" from our bodies' computer programs. Information technology specialists and quantum computer programmers (quantum computers likely would be necessary because of the massive complexity of the human body) might then have a lead role in health care, directing the work of doctors and other health care providers. Great achievements in health care might be achieved, including the possible eradication of disease.

Question 3. Since the everyday world we experience is *all* in our minds, our brain and mind must in some manner assemble this reality that we experience. Mental illnesses seems frequently to be associated with struggles to define reality. Is there perhaps a relationship between mental illness and this virtual imagery gone awry?

One possible relationship centers on the suggestion, as explained in the chapter above, that most people probably do not live in one alternative history but in several integrated *closely related* histories giving them the expansive and slightly ambiguous world most of us live in. Perhaps some people, due to failures of processing structures in the brain, either continuously or occasionally live in more than one *unrelated*, to them "real" everyday world, or in more alternative histories than the brain can effectively handle. The delusionary aspects of schizophrenia, a mentality characterized by inconsistent or contradictory elements,[102] may be an example of the way such a condition might be manifested. As many as 70 percent of schizophrenics hear voices, while a lesser number have visual hallucinations.[103] What to other persons are imaginary voices and hallucinations, to the person with schizophrenia are as real as any other experience.

Alternatively, other persons might experience too *few* alternative histories, causing them to live narrowly focused lives, preventing balance in their lives. This will bring to mind, for many people, autism, which is typically characterized by an extremely narrow focus on self[104] and great difficulty in communicating and forming relationships with other people.[105] As suggested above, so-called

---

102 *The New Oxford American Dictionary.*

103 See schizophrenia.com/earlysigns.htm.

104 See Simon Baron-Cohen, "Autism—'Autos': Literally, a Total Focus on the Self?" in T. E. Feinberg and J. P. Keenan, *The Lost Self: Pathologies of the Brain and Identity* (Cambridge, UK: Oxford University Press, 2005).

105 *The New Oxford American Dictionary.*

"normals" probably live in expansive and slightly ambiguous worlds of several overlapping, integrated, and closely related histories; this may allow them to become aware of the consequences associated with not only the centrally-focused paths of their histories but also of other paths in the cluster of closely related alternative histories. As a result, most nonautistic persons would become proficient and flexible decision makers, able to (and for children, demanding to) select from an ever-wider range of choices as they move through life. Autistic children and adults, if the hypothesis is correct that they experience too *few* alternative histories, would go through their daily lives without the learning experiences of their nonautistic peers, focusing narrowly on a context-free range of previous experience without reason or ability to do otherwise.

Question 4. The text in the chapter above states that the most important implication of the information-world postulate—that our world is made of information, not matter—is for the way we understand entropy. How, then, should we understand entropy?

Entropy is defined in its more general form as the degree of disorder or randomness in a system. In physics, entropy has been defined as a thermodynamic quantity representing the increased unavailability of a system's thermal energy for conversion into mechanical work. This definition is quite appropriate for a world made of matter. If, however, we consider the world at a more basic level to be made of information, then entropy, seen from this perspective, likely takes the form of increasing complexity. In either form, the arrow of time is inescapably associated with increasing entropy.

David Deutsch sees unavoidable increase in complexity to be a result of the increasing differentiation of the multiverse of which we are a part. New universes emerge only with change and this change inevitably brings added complexity.[106] Our attempts to cope with this complexity—to maintain order so that the change improves or at least does not lower the quality of our lives—typically involves technological change, which in turn produces more change and complexity. Seen from this perspective, we seem to be in a desperate—perhaps futile— race to deal with the increasing complexity of our world through technological change. Nevertheless, if we are to deal with what from this perspective seems to be a runaway increase in entropy, it is at this more fundamental level that we should focus our attention.

Change frequently is presented as progress—as an unalloyed good or a necessary evil. But sometimes change is followed by enormously complex problems that we are not prepared to handle. Development of the the atomic bomb ushered in the atomic age and, as former president Eisenhower pointed out in his farewell speech of 1961[107], the awesome destructive potential of atomic weapons led many nations to feel that a continual state of readiness to respond was a necessary defensive measure to protect their citizens from atomic attacks by other nations. This required the maintenance of standing armies and large defense industries with consequent disruptions to the economies and political structures of nations.

---

106 Interview with David Deutsch, in P. C. W. Davies and J. R. Brown, *The Ghost in the Atom: A Discussion of the Mysteries of Quantum Physics* (Cambridge, Eng.: Cambridge University Press, 1986), 86. See also Davies, *The Cosmic Blueprint*, 21.

107 Military-Industrial Complex Speech, Dwight D. Eisenhower, 1961, *Public Papers of the Presidents, Dwight D. Eisenhower*, 1960, p. 1035- 1040.

How do we avoid similar mistakes in the future? Some changes—and the change discussed in the previous paragraph may be an example—are probably unavoidable. Other changes might cause fewer problems for technologies to solve if we identified in advance those changes that will likely present problems and required that they conform to certain standards in the way they are introduced. But the breakneck rate of current change makes it difficult to see in advance of their introduction which changes will result in problems and what the problems might be.

I liken this inability to see the future to driving at night on a country road. We are driving too fast for the reach of our headlights. If we slow down and allow the road to be bathed in the glare of our headlights as we advance, the ability to see and avoid problems would be greatly enhanced. This more measured pace might be accomplished by appropriate regulation of free enterprise systems and by maintaining effective civilian control of powerful defense departments.

We would do well to heed the sage advice of Fred Rogers, host for many years of television's Mister Rogers' Neighborhood, who told us that we should all slow down, that life should be lived at a pace that allows us to discover that "deep and simple" is better than "shallow and complex." [108] While we cannot change the arrow of time—change will and must happen; perhaps we can slow the rate at which complexity increases.

---

[108] See the 2011 DVD, "Mister Rogers and Me."

CHAPTER 2
AT THE EDGE OF TIME

And again, again, I turned into the street, finding the
place where corners meet, turning to look again
to see where Time had gone. And all was there as
it had always been. And all was gone, and never
would come back again. (Thomas Wolfe)[1]

---

[1] Thomas Wolfe, ed. James W. Clark, Jr., *The Lost Boy; A Novella* (Chapel Hill,
NC: University of North Carolina Press, 1992), 76.

No man ever steps in the same river twice, for it's
not the same river and he's not the same man.
(Heraclitus)[2]

Given the nature of reality presented in Chapter 1 above, the
task of this chapter is to explain the role of time in our everyday lives. The central points of Chapter 1 of relevance for this
chapter are:

1. Time, as I will discuss it here, is a phenomenon of our
world (including the many worlds of the multiverse) alone; it
becomes a quite different concept outside that world.

2. Because the organizing principle used by the mind to assemble and to present to us the script for our everyday world
appears to be the very rules of logic and mathematics we use
in inductive reasoning, time as a process behaves in logical
ways in our everyday lives; there is nothing illogical about the
manner in which time unfolds. For brevity, I have called this
characteristic of our everyday world the "logical imperative."

3. Since elementary particles are the building blocks of our
everyday world, we may use quantum findings to understand
how time works—what I have called the "pointillist principle."

4. The things of our everyday world are abstractions and
have no existence apart from our minds. Our world is a
virtual reality.

---

[2] Heraclitis, trans. Brooks Haxton, Heraclitis Fragments (New York, NY:
Viking Penguin, 2003), 96.

What Is Time?

Consistent with the above foundational points, let me first offer a somewhat truistic definition of time. *Time is a medium of change*; it is the way we measure the relative rate of change of processes in our everyday lives. If nothing in our world (including our bodies) changed relative to anything else, time would not have meaning for us—the idea of time would not exist. (And, of course, life also would not exist, but that's another story.) As soon, however, as something in our world changes, the interconnectedness of everything in our world means that we must be concerned about the rate of change of one thing relative to another and out of this concern, time as a concept appears.

This truistic definition of time is important and is one that will appear repeatedly in our discussion below; however, it does not do much to help us to understand what time is. Let's turn our attention to this latter task.

While all of the above foundational points are important for this chapter, the most important of these is the last: our world is a virtual reality. How is that virtual reality presented to us?. Answering this question provides a context for our discussion of what time is?

Our virtual reality likely is presented to us as frames of a holographic movie. While this idea occurred to me as the outcome of a thought experiment, several other writers also have proposed variations of this same thought. Several researchers have proposed either that time might occur in discrete quanta (irreducible units) called "chronons" or that space might occur in discrete quanta called "hodons."[3] And Paul Davies has suggested that these quantized units might take the form of frames of a movie.[4]

A holographic projection has been suggested by Brian Greene as a way to understand the holographic nature of the universe. Greene says, "reality—not its mere shadow—may take place on a distant boundary surface, while everything we witness in the three common spatial dimensions is a projection of that faraway unfolding. Reality, that is, may be akin to a hologram. Or, really, a holographic movie."[5] And Jacob Bekenstein, one of the scientists who contributed to theoretical studies leading to the holographic-universe conjecture, says "our seemingly three-dimensional universe could be completely equivalent to alternative quantum fields and physical laws "painted" on a distant, vast surface.[6] As the reader will know, I have suggested that this projection occurs, not from a distant surface but from within our minds itself, from a "universe of the mind."

If our world is presented to us as frames of a movie, how are we to understand time? Let's go, in our imagination, to the edge of time and observe time as it happens—what would we "see?"

---

[3] See Henry Margenau, *The Nature of Physical Reality: A Philosophy of Modern Physics* (New York, NY: McGraw-Hill, 1950), 155-159.

[4] Paul Davies, "The Mysterious Flow of Time," *Scientific American*, September 2002, 32-37, esp.34.

[5] Brian Greene, *The Hidden Reality: Parallel Universes and the Deep Laws of the Cosmos* (New York, NY: Alfred A. Knopf, 2011), 238.

[6] Jacob D. Bekenstein, "Information in the Holographic Universe: Overview—The World as a Hologram," *Scientific American*, July 14, 2003.

We would see at a brief instant of time a film-like information structure that contains one slice in time of our everyday world. When I say film-like, I am saying that the "thickness" of this structure is an undetermined amount, but likely on the order of the Planck length. The Planck length (named after Max Planck, the originator of quantum theory) is a distance of less than 10^{-33} cm, commonly associated with Planck time, the time it would take light to travel across the Planck length, about 10^{-43} seconds. (Note that this fundamental unit of time is defined by reference to spatial distance—we will return to this point later.) Planck time is considered to be the shortest meaningful period of time; any two events that are separated by less than this amount of time can be considered simultaneous.[7] Since we live in a quantum world (in which energy exists only as discrete quantities), our moving present frame of time presumably must have a thickness on the order of one Planck unit for the particles comprising our world to be observed (and thus to come into existence).

If our world is presented to us as frames of a movie, we may think of this slice of our universe as one frame in a film or motion picture employing holographic film which captures a three-dimensional image[8] on a flat two-dimensional film, similar to the familiar

---

[7] David Darling, *The Universal Book of Mathematics: From Abracadabra to Zeno's Paradoxes* (Google eBook) (New York, NY: John Wiley & Sons, Aug 11, 2004), 245.

[8] It is likely that this film displays our world in more than three dimensions though we normally are aware of only three. See the answer to Question 1 at the end of this chapter.

bank credit card.[9] (While holographic film is not now commercially available, a working prototype has been developed and we appear, as of this writing, to be quite close to a commercially available product.[10])

To return now to our holographic film analogy of the way time passes in our everyday world, as readers know, a film is a series of images made up of individual images called frames. When these images are shown rapidly in succession, a viewer has the illusion that motion is occurring within the scene that was filmed. The viewer cannot see the flickering between frames due to an effect known as persistence of vision, whereby the eye retains a visual image for a fraction of a second after the source has been removed.[11]

Now imagine that the information structure that contains one slice in time of our everyday world is replaced by other frames sequentially so that we experience movement through our changing, scripted world one "frame" at a time just as the frames of a film capture virtual moments in the narrative related by the film. These frames are displayed at a frame rate equivalent to moving through space at the speed of light—186,282 miles per second, usually designated as constant "c" to indicate that this is the speed of light in a vacuum. This frame rate equivalent

---

[9] For an excellent technical description of a holographic image, see "Holography," Wikipedia, January 13, 2012.

[10] Larry Greenemeier, "Holographic Film for 3-D, sans Those Silly Specs," *Scientific American*, March 3, 2008.

[11] For a description of the art of movie making, see "Film Stock," Wikipedia, January 16, 2012.

to moving through space at the speed of light may be said to be the "fundamental frame rate" for the script of our everyday lives. This "fundamental frame rate" is an *absolute time* and an *absolute rest frame that provides us with a fundamental unit of time against which all other times in our everyday world can be measured.* (This is in contrast to Einstein's relativity theory, which allows neither of these foundational concepts.)[12]

Now that we know what time is, let's look at two implications of this understanding of time. I will group these implications under the following self-explanatory headings: Continual Creation and Personal Time.

Continual Creation

You probably noticed that the above representation of time, with time occurring in sequential frames of information, requires that our universe be *re-created within each frame.* This continual creation is what John Wheeler felt his delayed-choice experiment had demonstrated. Wheeler's hunch when he proposed this experiment, a hunch later confirmed by experiment, was that the universe is built like an enormous feedback loop, a loop in which we, by observing the universe, contribute to the *ongoing creation* of not just the present and the future but the past as well.[13] Presenting our universe to us as sequential frames of information means that there is no sequencing of events across adjacent frames; *each frame*

---

12 Stephen W. Hawking, *A Brief History of Time: From the Big Bang to Black Holes* (New York, NY: Bantam Pr., 1988), 33.

13 Folger, interview with John Wheeler. We *create* (present tense) the past? This part of John Wheeler's statement has not received the attention it deserves.

is discrete. Creation itself is therefore not a one-time process but is continual. This hypothesis, if correct, has several important corollary logical consequences.

We Are Moving through Space Not Time

A first logical consequence of the premise of continual creation is that *time is a measure of our rate of movement through (a changing) space.* To illustrate this point using our movie analogy again, assume that we are filming a train traveling at only five miles per hour (to avoid problems of moving the camera to follow the train) from left to right in our field of vision along a track in a wooded area of the United States. In an initial frame, the engine of the moving train is directly in front of a large oak tree on a hillside behind the track and a deer standing under the tree is feeding on a low-growing plant. In the next frame the scene may appear identical but an incredible number of changes will have occurred that are too small for us to notice, such as a leaf dropping from the oak tree. *One hundred* frames later (approximately four seconds later at an assumed frame rate of twenty-four frames per second), however, changes are significant enough for us to notice; for example, the first car behind the engine is now in front of the oak tree and the deer has been startled by someone waving from the train and has raised its head. Since we can only observe one frame at a time, the train, deer, and background items shown in the initial frame *no longer exist* for us; what exists is *only* this new version of the train, oak tree, deer, and background items.

Similarly, in the "movie" of everyday life, our present is always a *newly created* version of space. We, our virtual images, literally are moving through space at the speed of light, a speed of approximately 186,282 miles per second. We know this because for an object moving at the speed of light, time stops.[14] (We are normally unaware of this movement through space for literally *all time and space* is moving at a uniform rate; we have neither an inertial effect nor a reference against which to notice the movement.) This insight may be used to explain gravity, inertia, and a host of other physical phenomena. I will discuss some possible implications of the continual creation hypothesis for the way we view some of these topics in questions at the end of this chapter and in later chapters of the book.

A Cosmic Blueprint for Our Lives

Another logical consequence of this premise of continual creation is that there can be no causal relationship between frames. (For those concerned about the principle of local causation and Bell's Theorem, see Chapter 3 below.) Stated another way, the conditions of frame X can have no direct influence on what occurs in the next frame in the sequence, frame Y. Yet we know that our lives are not fragmented across time, that there is a carryover from one time period to the next. A likely explanation is that our universe is following a "cosmic blueprint" as it unfolds, as Paul Davies has suggested in his excellent book by the same name.[15]

If our lives are following a cosmic blueprint, does this mean that we live in a world in which all events, including human actions,

---

[14] John Gribbin, *Schrödinger's Kittens and the Search for Reality* (Boston, MA: Little, Brown and Company, 1995), 79.

[15] Paul Davies, *The Cosmic Blueprint: New Discoveries in Nature's Creative Ability to Order the Universe* (New York, NY: Simon and Schuster, 1988).

are ultimately determined by causes external to the will? Notice that this question, in our world of incomplete knowledge, can only be answered in relativistic terms: for example, free will obviously will mean different things for stones, amoebas, and animals. Given this caveat of relativity, the question of interest to most of us when the issue of free will is raised is this: *has the script for the lives of human beings been written in advance?* The answer seems clearly to be: it has not. Indeterminism manifests itself most conspicuously on an atomic scale of size where the Heisenberg uncertainty principle dictates that the observable properties that characterize a physical system are undecided from one moment to the next.[16] Applying the pointillist principle, this indeterminism must also exist at the macro level of our everyday lives.

And at the macro level, it is obvious that differences in our lives *do* depend on choices made. Chapter 1 presented the idea that each of us at every moment in time has choices in part determined by our past histories that will in part determine our future histories. These choices and their past and future alternative histories are logically and probabilistically related to our current history. Our cosmic blueprint unfolds in logical ways (point 3 above) in response to the choices we make in life. Laurence LeShan and Henry Margenau make a similar argument.

> Our thesis is that quantum mechanics leaves our body, our brain, at any moment in a state with numerous … possible futures, each with a predetermined probability. Freedom involves two components: chance (existence of a genuine set of alternatives) and choice. Quantum mechanics

---

16 See Davies, "The Mysterious Flow of Time," 37.

provides the chance, and we shall argue that only the mind can make the choice by selecting ... among the possible future courses."[17]

Personal Time

You probably also noticed that, if the script for our lives originates in and is transmitted through each of our minds, the time our universe experiences is, for each of us, our *personal time*. One way that time for you is personal is well-established by scientific evidence. If you want to have your time proceed more slowly than the time of other persons accompanying you on your journey through life, you can, in theory at least, move very rapidly.

The well-known story of the twin brothers separated by space travel illustrates this point. In my version of this story, the twins meet at a local airport that serves as a hub for flights to outer space where one of the twins, Mack, will begin a journey. Mack explains to his twin brother, Mick, that he expects to be away on business for about a year. He then boards a space shuttle that flies at a significant proportion of the speed of light to a distant planet outside our solar system. After a year, according to his calendar watch, Mack lands at the same airport and meets Mick, who is now fifty years older than Mack!

Note several important points about the nature of time illustrated by the twin scenario described here: First, note that when Mack returns to Earth, everything about Mack's world while in space, both those things at the macro and those at the quantum

---

[17] Laurence LeShan and Henry Margenau, *Einstein's Space and Van Gogh's Sky* (New York, NY: Macmillan, 1982), 240:

levels—everything that moved through space with him—is fifty years younger than Mick and his world.

Second, note that neither twin noticed anything different about time's passage while they were separated; if asked, each would have said, "I noticed nothing different about the rate at which time passed."

Third, note that, in the dual-reality framework, the only thing about Mack's trip as described here that matters *insofar as the relationship of Mack's time to that of Mick* is that Mack moves while Mick does not move—each referenced by his initial position at the time of separation. It is this mobility difference that produces the difference in the rate at which each twin ages. To illustrate this, at the beginning of Mack's journey, as his ship picks up speed, the difference between the rate of passage of Mack's time and the rate of passage of Mick's time steadily increases; at the end of his journey, as his ship slows to land at the spaceport, the difference between the rate of passage of Mack's time and Mick's time steadily decreases until there is no difference when his ship lands. If, for a period of time during Mack's journey his ship stops moving, the rate of passage of time in each twin's world would be the same (ignoring the relatively small gravitationally-induced differences in time passage for the twins) during that period of time. Mack's space voyage changed the rate of passage of his own time; it had no effect in any sense on the rate of passage of Mick's time.[18]

---

[18] The age difference between the twins also may be explained using inertial frames as, for example, in Paul G. Hewitt, *Conceptual Physics*, 8th ed. (Reading, MA: Addison-Wesley, 1989), 45-51.

Fourth, note that in the framework of the holographic movie model of time we are only able, by our movements, to slow time; time cannot be made to go faster. We can move toward our personal projector but not away from it. (This point will be explained more fully in Chapter 4.) If you are not sure which twin was the traveler, compare their ages. The one who aged the least during their separation is the one that traveled.

Finally, note that in this example, there are two measures of the rates of the passage of time for each twin, one measured by his own clock and another measured by his brother's clock. Let's call these two times, respectively, real time and comparative time. This final note is vital to understanding the nature of time and therefore deserves more discussion.

Real Time and Comparative Time

To bring into sharp focus the difference between these two measures of time, assume that Mack and Mick both work for the same employer and that Mack begins his business trip when the twins are only twenty years old. Mack is moving in relation to his own personal "projector," which produces his personal world and time, and which "runs" at a frame rate exactly equal to the projector creating Mick's world. By moving toward his projector, Mack is slowing down his rate of movement through time (as compared with Mick) and thus Mack will appear in many fewer frames of his projector, during his journey, than the nonmoving Mick will appear in his projector. The importance of this difference in number of frames experienced is that *change in our universes can occur only as we move between the frames displaying our life stories.* Therefore, when Mack returns from his travels, at age twenty-one, he can present only one year's worth of work to his employer (remember, he *was* on a business trip!). Brother Mick, meanwhile, at age seventy-one, has worked for their common employer for fifty-one years while Mack was away and has a corresponding amount of work to show for it. In every sense, Mack has had, during his travels, only one year of life to enjoy, while Mick has had fifty-one years of life to enjoy. If each twin lives to his normal life expectancy, and if Mick is a guide to how long Mack may be expected to live, Mack still has at least fifty more years of life (and employment) ahead of him, and at that time he will have had exactly the same life span (past frames of time) as Mick had fifty years ago.

It should now be apparent why we call time as measured by Mack on his calendar watch Mack's "real time" for it is a reliable measure of the rate at which he is aging. Similarly, it is appropriate to call the time measured on Mick's calendar watch his "real time" for again it is a reliable measure of the aging process of his life.

But there also is comparative time. Mack may well insist that he has been away only one year and Mick may well insist that Mack was away fifty-one years. Which twin is correct? Each of us has an inherent tendency to assume that the way we see the world is the way others must see it—I taste a food and insist that you must also experience the same taste sensation as I. But that will not necessarily happen since each of our taste sensations are entirely within our minds. Similarly, if we ask which twin is correct about the length of time they were apart, the answer, of course, is both. Time for each twin is personal and comparative time serves the purpose of allowing predictions and coordinated effort.

Does all of this really matter? Yes, because it helps us to understand the limits of our ability to manage time. We can and do change the rate at which we pass through time *as compared with other parts of our universe* but, to this point, we are unable to change the fundamental frame count that determines our age. Mack's only advantage gained by aging comparatively more slowly than his brother, if it may be considered an advantage, is that he will be able to live within a different (future) world with a different cohort of people than his brother's cohort. He could, for example, bring to those future people actual memories of a somewhat distant past. (And would these future people believe Mack to be a "time traveler"?)

While we cannot propel macro-sized objects, including humans, at a significant portion of the speed of light, our modern atomic clocks allow us to confirm that even at the relatively slow velocities of modern aircraft, time does move more slowly for the aircraft and its contents, including its passengers, than for non-travelers on the ground. Also, for reasons that we will discuss in Chapter 4, atomic clocks at different altitudes (and thus at different gravitational potentials) will eventually show different times. These points confirm that, while all our personal (real) times have a commonality in that we move through changing space (space changes for us) at the same rate, each of us has a personal comparative time that may differ from that of others depending on whether we are flying in a plane or living at a high altitude. While these effects are extremely small, they are there and must be accounted for in any theory of time. *We each have our own, independently determined, personal comparative time.*

Does Time Dilate for Moving Non-Living Things?

Yes. The space-traveling twin saga makes this clear. Mack, the space traveling twin, and everything that moved through space with him aged more slowly than the things of Mick's earth-bound world. Both those things at the macro level, such as the space ship in which Mack traveled, and those at the quantum level, such as the elementary particles of which the ship is "composed", when traveling at a high speed, are subject to a comparative time distortion as was Mack.

Muons are quite interesting in this regard. Muons are elementary particles formed when the earth's atmosphere is bombarded by particles from space called cosmic rays. As these cosmic rays travel toward Earth, they interact with atoms high in the atmosphere, often producing showers of muons. Even though muons travel at a large fraction of the speed of light, because they have a very short lifetime (only a couple of microseconds before they 'decay' into other kinds of particles), they do not live long enough to travel through our atmosphere and reach the surface of the earth. Yet, from our perspective, most do reach the ground! This is because they are traveling so fast toward a nonmoving earth (that is, the muons are moving toward the earth, not the reverse, each referenced by its position at the start of the muon journey) that time slows markedly for them (they experience many fewer frames of time) *as compared* with Earth and its inhabitants. *From the standpoint of an Earth-based laboratory scientist* who detects the muons, they appear to live about nine times longer than they would at rest. *From the standpoint of the muons* (if they had standpoints and carried watches!), their lifetimes would still be a couple of microseconds, but they would share that lifetime with several generations of their stationary brethren and with laboratory scientists who observe them on Earth.[19]

It is well-established, then, that comparative times differ somewhat for all of us—we each have our own personal time. Is it possible that our personal times differ more and more often than we realize? There are several instances reported in the literature and anecdotally where personal times of ordinary people are believed to differ markedly from other persons with whom they are interacting. Let's examine the evidence for this.

Comparative Time Differences in Infrequent Circumstances

Those instances of comparative time differences found by the author apparently occurred because of what the people whose comparative times differed were observing at a given time—upon their focus of attention. Exactly how that may occur will be discussed later in this chapter but intense focus by an individual upon other persons or things appears to be a key requirement in each of these instances for a comparative time difference to occur. Instances of possible comparative time differences found by the author are: when a person observes a near accident or impending peril, when a person has what has been called a "peak life experience", and when a person is under hypnotic trance.

---

[19] Gribbin, 78; Brian Greene, *The Elegant Universe: Superstrings, Hidden Dimensions, and the Quest for the Ultimate Theory* (New York, NY: W. W. Norton and Company, 1999), 42.

Near Accident or Impending Peril

It is a somewhat common experience for persons involved in an accident or near accident or other dangerous event to feel that time slows dramatically during the event.[20] Intense focus on what is happening is, of course, an inevitable outcome for most of us in such circumstances; in effect, the total world of our awareness becomes what is happening in front of us. And this focus may cause the *activities associated with the peril to seem to slow* for some observers, allowing them more time to consider and react to the danger and may explain why some observers at an accident scene are able to take appropriate action at the time while others only later realize what they should have done.

This response clearly would have survival value and may have evolved to allow humans (and perhaps our fellow nonhuman creatures as well) to deal with an often threatening world. Those people for whom time seems to slow during a time of danger literally may have more time in which to think and to act than do persons who do not experience this time distortion response.

Peak Experience

Another circumstance where people frequently report a slowing of time is when they have a "peak experience." Abraham Maslow, humanistic psychologist who in the 1950s popularized the

---

[20] Peter Ulric Tse, James Intriligator, Josée Rivest, and Patrick Cavanagh, "Attention and the Subjective Expansion of Time," *Perception & Psychophysics*, 2004, 66 (7), 1171-1189, esp.1171; Linn F. Cooper, MD, and Milton H. Erickson, MA, MD, *Time Distortion in Hypnosis: An Experimental and Clinical Investigation* (Boca Raton, Fl.: OTC Publishing, 2004), 29.

concept of hierarchically arranged needs, first used the term peak experience to describe certain transpersonal and ecstatic states, particularly ones tinged with themes of euphoria, harmonization, and interconnectedness, which infrequently occur in the lives of people but may affect the overall direction of their lives thereafter. According to Maslow, all the peak experiences he studied involved

> ... a very characteristic disorientation in time and space. This goes so far that it would be more accurate to say that in these moments the subject is outside time and space subjectively.[21]

Later studies of peak experiences in various occupations, notably those of performers such as professional football players, race-car drivers, or runners, confirmed that most of these peak experiences involved what Maslow called disorientation in time and space.[22]

As pointed out in Chapter 1, performance of a physical activity, by those who do it well, requires, in part, concentration, freedom from distraction, and sustained alertness. Extraordinary performance depends on one's ability to focus unbroken attention on the *space, objects, and people involved as determinants of one's competitive performance, and on one's own kinesthetic sense of the body.* These elements determinative of performance typically are learned by countless hours of practice. Focus of attention upon these elements by the

---

[21] A.H. Maslow, "The Good Life of the Self-Actualizing Person," in Theron M. Covin, *Readings in Human Development: A Humanistic Approach* (Beverly Hills, CA: MSS Publishing Corp., 1974), 46-53.

[22] Kenneth Ravizza, "Peak Experiences in Sport," *Journal of Humanistic Psychology*, 17: 35-40, 1977, reprinted in *Essential Readings in Sport and Exercise Psychology*, Daniel Smith and Michael Bar-Eli (eds.) (Champaign, IL: Human Kinetics, 2007), 122-125, esp. 123.

great performers at critical times in their performance, according to their self-reports, frequently leads to an apparent slowing of time for them.

Three typical examples illustrate the time-distortion experience as great performers frequently see it. In American football, John Brodie, former quarterback for the San Francisco 49ers, says that in the most intense moments of a football game

> … time seems to slow way down, in an uncanny way, as if everyone were moving in slow motion. It seems as if I had all the time in the world to watch the receivers run their patterns, and yet I know the defensive line is coming at me just as fast as ever. I know perfectly well how hard and fast those guys are coming and yet the whole thing seems like a movie or a dance in slow motion. [23]

Former Grand Prix champion race car driver Jackie Stewart believes the most important requirement for successful performance of his sport is to synchronize himself with the elements he's competing against, the motor car, and the track.

> Your mind must take these elements and completely digest them so as to bring the whole vision into slow motion. For instance, as you arrive at the Masta [a particularly frightening corner on the Formula 1 circuit] you're doing a hundred and ninety-five mph. The corner can be taken at a hundred and seventy-three mph. At a hundred and ninety-five mph you should still have a very clear vision, almost in slow

---

[23] Michael Murphy and John Brodie, "I Experience a Kind of Clarity," *Intellectual Digest*, vol. 3, no. 5 (January 1973), 19-22, quoted in Murphy and White, 46.

motion, of going through that corner — so that you have time to brake, time to line the car up, time to recognize the amount of drift, and then you've hit the apex, given it a bit of a tweak, hit the exit and are out at a hundred and seventy-three mph. [24]

Runner Steve Williams, former 100-meter dash champion, says "if you do a 100 right ... that 10 seconds seems like 60 . . . . Time switches to slow motion."[25]

Note that it has not been said that practice and innate ability are unimportant. And none of the performers claimed to have been given extraordinary powers or strength. They only contended that, for them, time slowed down and afforded them the opportunity to better apply their innate abilities and carefully honed skills.

Other performers among the many who have reported the sensation of time slowing down during peak performances are ballet dancers, basketball players, and golfers.[26]

Hypnotic Trance

Unlike the examples of time distortion presented above, of individuals experiencing impending peril or a peak performance event, where the individuals experienced a sensation of time slowing, individuals in a hypnotic trance, in the experiments to be described below, feel that time passes for them *at its usual rate* while in

[24] Peter Manso, *Vroom! Conversations with Grand Prix Champions* (New York, NY: Funk and Wagnalls, 1969): 180-181).

[25] Kenny Moore, "A Night for Stars, Both Born and Reborn," *Sports Illustrated*, vol. 46 (May 23, 1977), 32-34, esp. 34.

[26] See Murphy and White, various pages.

the trance. This difference is instructive of the difference between the mechanism causing time distortion for individuals experiencing impending peril or a peak performance and individuals in a hypnotic trance. We will return to this point below.

The most extensive series of experiments examining time distortion during hypnosis was conducted by Dr. Linn F. Cooper.[27] Recall from the discussion in Chapter 1 of hypnotism used to enhance creativity that Dr. Cooper's instructions to subjects were as follows: "Now, give me your attention, please. When I give you the starting signal, by saying 'Now', you're going to (you will) …" followed by instructions about the experience the subject was to have. At the end of a predetermined time, Dr. Cooper would halt the trance by giving a prearranged termination signal, typically the word "blank," meaning the subject's mind was to become blank at that point.

It was during these experiments that Dr. Cooper found, to his surprise, that time distortion was occurring[28] and that time distortion was aided by assuring the subject that he or she would have all the time he or she needed and would not have to hurry.

We use such suggestions as the following:
You will not have to hurry, for you will have all the time you need.

---

[27] Linn F. Cooper and Milton H. Erickson, *Time Distortion in Hypnosis: An Experimental and Clinical Investigation* (Boca Raton, FL: OTC Publishing Corp, 2004). (First published 1954). (The book is divided into three parts; Dr. Cooper's research is reported in Part I.)

[28] See Cooper and Erickson, 39-41, for Dr. Cooper's remarkable account of this discovery.

Don't hurry; take your time.

Remember, you have an unlimited amount of special time
available and will take as much of this as you need to finish
the task without hurrying.[29]

The dress-design experiment conducted by Dr. Cooper and described in Chapter 1—presented there to illustrate how hypnotism
might be used to enhance creativity—illustrates as well the manner
in which time distortion frequently occurred during hypnosis.

Recall that the subject in the experiment performed six tasks,
each a different, original dress design. The subject was waked immediately after the completion of each trance activity and (with
the exception of Task 1, where she was only asked to describe the
design) was asked to draw a picture of the dress she had designed
and to describe it briefly. These pictures and her descriptions of
the design were quite detailed.[30] Recall also that the subject was a
skillful designer, not a novice; as such, she would seem well quali-
fied to estimate, within an acceptable margin of error, the time she
spent on each design. Her estimates of the time required for her to
design, in the trance state, each dress and the time actually allowed
for the experiment are as follows:

Task	Subject's estimate of time spent	Clock time actually allowed
First Design	1 hour	10 seconds
Second Design	1 hour	10 seconds

---

[29] Cooper and Erickson, 48.

[30] For drawings and descriptions of the dresses, see Cooper and Erickson,
139-143.

Third Design	several hours	5 seconds
Fourth Design	1 1/2 hours	≤ 1 second
Fifth Design	1 1/2 hours	≤ 1 second
Sixth Design	1 hour	10 seconds

The differences in the subject's estimate of time spent on the task and the amount of clock time allowed for the experiment are truly remarkable. As indicated in the data above, Dr. Cooper found that, in general, there was little if any connection between the time he allotted for an experiment and the rate of passage of personal time of the subject.[31] The rate of passage of a subject's time appeared to be fully under the subject's control—it was his/her personal time.

(It may help to understand what appears to be happening in this experiment if the reader compares this experiment with the space-traveling twin saga—think of the hypnotist as Mack, the space-traveling twin, and the subject as Mick, the stay-at-home twin.)

Apparently, no attempt to repeat Dr. Cooper's experiments has been conducted. (The Zimbardo, Marshall, and Maslach experiment discussed below does not cite Dr. Cooper's experiments.) It is critical that an attempt to confirm Dr. Cooper's results duplicate the procedure Dr. Cooper used since the extreme suggestibility of the hypnotic subject makes the procedural relationship between hypnotist and subject a critical variable of the study.

---

[31] For a discussion of this point, see Cooper and Erickson, 67-77.

However, later experimental studies of hypnosis have consistently shown alterations in hypnotic subjects' sense of time. These alterations were typically considered to be sensory anomalies and were variously called: "expansion of time," "time slowed down," and "an altered sense of time."[32] Two frequently cited explanations for the anomalies are the "busy beaver" hypothesis, which posits that demands placed on attentional resources during hypnosis caused faulty estimates of clock time intervals[33] and the "tardy time keeper" hypothesis, which suggests that time judgments are derived from an internal clock, which runs slowly during hypnosis.[34] As might be supposed, with one exception, none of the studies attributed the anomalies to an actual change in the rate at which time passed for the subject. The exception was a study by Philip G. Zimbardo, Gary Marshall, and Christina Maslach, in which hypnotized college students were given the suggestion to "allow the present to expand and the past and future to become distanced and insignificant. "The authors found "profound effects upon thinking, feeling, and acting" among hypnotized subjects, suggesting that they were indeed allowing the present to "expand" (as compared with control groups) in response to a mere verbal suggestion to re-create one's perceived boundaries between the present and past

[32] Etzel Cardeña, "Anomalous experiences and hypnosis," *Proceedings of the 49th Annual Conventions of the Parapsychological Association*, 2006 and 2007, 32-42.

[33] See R. St Jean, K. McInnis, L. Campbell-Mayne, and P. J. Swainson, "Hypnotic Underestimation of Time: the Busy Beaver Hypothesis," *Journal of Abnormal Psychology*, August 1994,103 (3), 565-9.

[34] See "Hypnotic Time Perception: Busy Beaver or Tardy Timekeeper?" P. L. Naish, *Contemporary Hypnosis*, 2001, 18, 87–99.

and future.[35] These findings of an "expanded present" clearly seem to be a confirmation of Dr. Cooper's results seen from a different perspective—that is, in Dr. Cooper's experiments, since the hypnotized subjects stated that they performed tasks that normally would require perhaps an hour of their time in perhaps two seconds of the hypnotist's time, the subjects could be seen as experiencing an "expanded present."

What Happens during Time Distortion?

Two influences may be at work causing the time distortion that seems to occur for ordinary people depending upon their focus of attention at a given time. As pointed out above and in Chapter 1, we bring our everyday world, our universe, into existence by becoming aware of it. And that universe, in its entirety, is moving through space at approximately the same rate. But, as we narrow our focus on our everyday world, two things may be happening that cause time distortion.

First, a "watched pot" effect my be occurring—we are slowing the passage of time for the things we observe simply by watching them. The "watched pot" experiment of quantum physics suggests how this time distortion may occur. This experiment has shown that elementary particles do not change their quantum state while being watched. John Gribbin describes the watched pot experiment as follows:

---

[35] P.G. Zimbardo, G. Marshall, and C. Maslach, "Liberating Behavior from Time-bound Control: Expanding the Present Through Hypnosis," *Journal of Applied Social Psychology*, 1971, 1, 4, 305-323

In some sense particles of matter, including atoms, do not really exist, as particles, when nobody is looking at them— when no experimental observations are occurring. A natural corollary of this idea is the idea that "a watched atom can never change in quantum state, as long as it is being watched." Even if you prepare the atom in some unstable, excited high-energy state ... , if you keep watching it the atom will stay in that state forever, trembling on the brink, but only able to jump down to a more stable lower-energy state when nobody is looking." A watched quantum pot, theory says, never boils. And experiments ... bear this out.[36]

Objects in our daily lives are in some sense collections of atoms. Consequently, consistent with the pointillist principle, there is no reason that macro-systems should be different. So we may be able to decrease the rate at which comparative time passes for objects in our environment by choosing to focus our concentrated attention upon those objects.

A second factor that may be causing time distortion in everyday life is that in the instances of time distortion discussed above, when we are narrowing the scope of our awareness of our world, this is causing a change in the gravitational effect on the rate of

---

[36] John Gribbin, *Schrödinger's Kittens and the Search for Reality* (Boston, MA: Little, Brown and Company, 1995), 132. See pp. 132-135 for a quite accessible description of the experiments demonstrating this phenomenon at the quantum level. Also see Charles Seife, *Decoding the Universe: How the New Science of Information is Explaining Everything in the Cosmos, from Our Brains to Black Holes*, (New York, NY: Penguin Group (USA) Inc., 2006), 197-199.

passage of comparative time for us. In the normal course of everyday life, our comparatively extensive scope of awareness of our world means our world is massive and the associated gravitational effect causes time to pass more slowly for us. On the other hand, a quite restricted scope of awareness, as during a time distortion event, causes our world to become less massive and the associated less powerful gravitational effect may allow our comparative time to pass more rapidly. (This point will be more fully developed in Chapter 4 below.)

Let's look at time distortion during impending peril and peak experience and during a hypnotic trance with these factors in mind.

Time Distortion during Impending Peril and Peak Experience

Time distortion while observing impending peril and while experiencing a peak performance may operate in similar fashion. By focusing narrowly on selected events, some observers of impending peril and some peak performers (let's call them collectively "participants") narrow their focus upon their world which, to the participants, *seemingly* slows their own passage through time.

I have said that time for participants *seems* to slow—but this may not be what actually happens. The participants' intense focus on, respectively, activities associated with a peril or elements determinative of performance, means that the world of the participants, *his or her entire field of awareness, becomes these events* unfolding before him or her. This intense focus on the events of a sharply compacted world *causes the events unfolding before the participant to slow down* dramatically while lessened gravitational effects allow time to pass *more rapidly* for the participants. Participants experiencing time distortion thus have more time to act during the event than they would otherwise have. If we acted on these assumptions and could have measured the aging process in both NFL quarterback John Brodie and the defensive players on the other team (see above), we would have found that during the particular play in which time distortion occurred, John Brodie would have aged unusually rapidly, much more rapidly than the defensive players, who would have aged at their "normal" rate.

By an intense act of will, thereby triggering a "watched-pot" effect and a gravitational effect, Brodie is able to "slow" temporarily the advance of the defensive linemen—but only temporarily. Eventually, when John Brodie must again focus upon his larger world, just as the "watched pot" of quantum physics must eventually boil when the watchful scientists terminate their experiment, the defensive linemen will no longer be slowed for Brodie.

Since the script of the life of John Brodie is generated within his mind along with his personal time, spectators had no direct way to see that the critical events they were observing, events that in world time required perhaps four seconds, were being experienced as perhaps twenty-four seconds by Brodie. What they *could* see was that John Brodie was making extraordinarily good use of the apparent time involved (four seconds in this example) relative to the elements he was competing against, and thus was accomplishing feats that NFL quarterbacks seldom accomplish.

Let's look now at time distortion during a hypnotic trance,

Time Distortion during a Hypnotic Trance

Before a subject is hypnotized, the hypnotist and the subject share a mutual awareness—they are quite aware of each other and thus the rates at which they move through space (their comparative times) are closely synchronized. During hypnosis, this synchronicity of world times is lost. The hypnotist directs the subject's thoughts, feelings, and behavior by instructing him or her to concentrate on particular images, ideas, or tasks. Subjects under hypnosis, as stated above, are far more open than usual to suggestions and are able to focus intensely on specific thoughts, feelings, sensations, or tasks. When the hypnotist leads the subject to focus attention on the performance of a task, this has the effect of bringing about what Milton Erickson, the premier hypnotism expert of the twentieth century, contended was a necessary and essential feature of a hyp-notic trance, *a total lack of awareness by the subject of his or her [laboratory] surroundings.*[37] Dr. Cooper agrees and refers to this phenomenon as the subject becoming detached from his/her surroundings in the physical world.[38]

A lack of awareness by the hypnotic subject of a larger world beyond task performance means that the subject is, literally, in a severely constricted world of his or her own that consists of his or her self and elements determinative of performance of an as-signed task in that world. *The rate of time passage in the entranced sub-ject's world now has nothing to do with time passage in the hypnotist's world and vice versa.* And without the slowing effect of a mutual awareness

---

[37] M.H. Erickson, "Deep hypnosis and its induction," in *Experimental Hypnosis,* L.M. LeCron (ed.) (New York, NY: Macmillan, 1952), 70-114.

[38] Cooper and Erickson, 56.

of the subject and the wider world that he or she and the hypnotist usually share, gravitational effects on comparative time for the subject are sharply reduced. Comparative time thus moves more rapidly both for the subject and for the elements determinative of performance in that world (such as any machinery needed for the task). The subject is able, as a consequence, to accomplish much more work than he or she could have accomplished in the larger world of the hypnotist. (Similarly, Mick was able to accomplish much more work during their separation than his space-traveling twin Mack in the the space-traveling twin saga discussed earlier.)

Time distortion during a hypnotic trance where the entranced subject experiences complete detachment from the larger world differs from time distortion during impending peril and peak experience where attachment to events occurring in the larger world of the participants is essential to the experience. Another difference between time distortion during hypnotic trance and that during impending peril and peak experience is that in the former there is no urgency to act on the part of the hypnotized subject. As stated above, Dr. Cooper found that time distortion was aided by assuring the subject that he or she would have all the time he or she needed and would not have to hurry. There is thus no watched-pot effect to cause comparative slowing of the events in the subject's world. Both hypnotist and hypnotized subjects therefore feel that time passes normally for them during the trance since everything in each of their worlds pass through time at the same rate. (Similarly, there was no sense that time passed in other than its usual fashion for each of the twins in the space-traveling twin saga discussed earlier.)

Can the Holographic Movie Model of Time Be Proven?

The model of time presented above might be termed a holographic-movie or motion-picture model of time since it defines time as the rate of change in our life stories presented to us as sequential frames, each depicting one scene in our life story similar to the frames of a motion picture. As with any hypothesis, proof, if it comes, will come in the form of successful application of its ideas. Beyond this, the model has two key features, continual creation and personal time; if these key features can be proven or disproven, this would go far to establish or refute the model. (For reasons discussed earlier in this chapter and having to do with limitations on our knowledge, ultimate proof or disproof is not possible.)

Continual Creation

Two approaches to proving or disproving (in the limited sense in which this is possible for us) continual creation seem feasible: finding an indivisible unit of space and demonstrating that each frame of the holographic movie is a new creation. Each of these approaches are discussed below.

Continual Creation and the Need for Indivisible Space

Creation likely is a one-time event if space is divisible to the vanishing point. A necessary feature of continual creation is an indivisible unit of space within which each frame of our time may be contained. (Note that while we often talk of an indivisible unit of time, time has no independent existence apart from space; time is defined by reference to distance and therefore it is space, not time, that must be indivisible.) If we find no indivisible unit of space, the model comes into question. It has not been established at the present moment whether space is quantized.[39] Is it likely that this can be done in the future?

Henry Margenau suggests that the hodon, his name for the indivisible unit of space which corresponds to what here is called the frames of the holographic movies of our lives, is about 10^{-13} cm and the corresponding chronon, the smallest interval of time, would be of the order of magnitude 10^{-24} seconds.[40] The shortest period of *time* that can currently be resolved is about 10^{-15} seconds. The Planck unit of time, the time it would take light to travel across the Planck length, is about 10^{-43} seconds. As stated above, Planck time is usually considered to be the shortest meaningful period of time. If we do not find an indivisible unit of space at Margenau's hodon length or the Planck length, the holographic movie model of time begins to come into question.

Scott Diddams and Tom O'Brian of the Time and Frequency Division of the National Institute of Standards and Technology, state: "… it generally takes more and more energy to probe physical

---

[39] "Is Time Quantized? In Other Words, Is there a Fundamental Unit of Time That Could Not Be Divided into a Briefer Unit?" *Scientific American*, Thursday, October 21, 1999, 12.

[40] Henry Margenau, *The Nature of Physical Reality*, 158,

events on shorter and shorter time scales. Probing the ... Planck length is ... far beyond the reach of current accelerators."[41]

If machines fail, perhaps theory can save the day. String theory is believed to offer a view of our everyday world that is equivalent to the view now provided by quantum mechanics.[42] String theory might therefore offer a way to determine whether space in our everyday world has an indivisible unit.

Each Frame of the Holographic Movie a New Creation

At the level of tiny particles, the laws of physics are symmetrical in time. A reaction that proceeds in one direction (such as particle A transforming into particle B) is just as likely to occur in the reverse direction (particle B transforming into particle A).[43] Experimental findings suggest, however, that there should be exceptions to this rule. In fact, the Standard Model of physics, the mathematical description of the elementary particles of matter and the electromagnetic, weak, and strong forces by which they interact, requires such an exception.[44] So too does the holograph-

---

[41] Answer to a question posed by a reader of *Scientific American*, posted at ScientificAmerican.com, December 27, 2004: "What is the fastest event (shortest time duration) that can be measured with today's technology, and how is this done?"

[42] Juan Maldacena, "The Illusion of Gravity," *Scientific American*, November 2005, 56-57.

[43] Steve Nadis, "Time Assymetrey Finally Found." *Discover*, June 2013, 14. Also see Quinn, 5.

[44] Helen R. Quinn, "Time Reversal Violation," Talk presented at the Discrete '08 Conference, and published in the Journal of Physics Conference Series, p. 3.

ic-movie model of time require an exception to this rule. This model assumes that our world is presented to us in frames, each frame a new representation of our universe—a new creation. And creation is, at its essence, symmetry-breaking in the truest form: *something from nothing is not a time reversible event.*

In the June 2013 issue of *Discover* magazine, Steve Nadis reports that scientists have finally discovered direct evidence of *a lack of symmetry in time* at the particle level, an instance of "time reversal violation." The discovery was made by studying the behavior of pairs of B mesons in superposition. Researchers exploited the fact that particles in superposition are anticorrelated, if you know the properties of one, you know the properties of its partner. Therefore, when one of the B mesons decays or changes form, physicists know its partner's state before it, too, decays. Researchers kept track of the transitions that the partner made and discovered that some reactions occur much more frequently with time going forward than they do with time in reverse.

This finding, if confirmed, may be seen as partial confirmation of the dual-reality hypothesis and the holographic-movie model of time.

Personal Time

The second feature of the hypothesis of a holographic-movie model of time on which the model stands or falls is personal time, so called because, according to this hypothesis, each of us has a unique rate at which things change in our everyday world. That each of us has a personal time follows logically from the hypothesis of Chapter 1 that we each live in a universe of our own making.

This feature, personal time, as such is obvious and needs no proof. Time can hardly be other than personal. Some of us travel a lot, others travel little, and we live at different altitudes; these influences on the rate at which time flows, though extremely small, do exist and *do change the rate at which time passes for each of us uniquely*. And it is well established that the twin paradox *could* occur if the technology permitted; each twin would experience real time as passing at a different rate. Thus (comparative) time *can* differ for each of us. (Note again that if time passes at different rates for each of us, then, as illustrated by the experience of the twins described above, we must also age at different rates. I will return to this point in a moment.)

While no proof seems necessary to prove that each of us has a personal time, it is a different matter to prove that we can, in effect, manage the rate of flow of our personal times by the way we behave in relation to events of our scripted personal world, as in the examples above of persons observing impending peril, experiencing a peak performance event, or being in a hypnotic trance. Can we collect tangible evidence other than the testimony of the participants themselves to demonstrate this point? Three possible approaches to collect such evidence are described below.

The SCAD Experiment

This experiment was an ingenious attempt to establish whether time "really" does pass differently for people who are experiencing impending peril. This experiment illuminates the difficulty of proving the reality of something which, to observers, is "all in the mind" of another person.

As described above, observers of (or participants in) a life-threatening event frequently report that time seemed to have moved in slow motion during the event. To determine whether time really did slow down in such circumstances, Stetson, Fiesta, and Eagleman had volunteers do a SCAD (Suspended Catch Air Device) dive in which participants experienced a 150-foot free fall for about three seconds before landing safely in a net. Each participant wore a small electronic device, called a perceptual chronometer, which, under normal circumstances, flashes numbers just a little too fast to be seen. Researchers assumed that if time slowed, the numbers could be read by participants.

There was no evidence that participants could read the numbers more accurately during the free fall than they had in control readings while on the ground. However, these results were in apparent conflict with the fact that participants retrospectively estimated that their own fall lasted 36 percent longer than the falls of other participants.[45]

Unfortunately, the SCAD experiment was not appropriately designed to test the "personal time" explanation for why time seems to slow (but actually flows more rapidly) for some observers of life-threatening events—this is understandable, of course, since that was not the purpose of the design. It was explained above that time flows more rapidly for observers of such an event because of *an intense focus on activities associated with the peril.* There are two problems with the design of the SCAD experiment. First, since participants are completely reliant

---

[45] C. Stetson, M. P. Fiesta, and D. M. Eagleman, "Does Time Really Slow Down during a Frightening Event? *PLoS ONE*, 2(12), 2007: e1295. doi:10.1371/journal.pone.0001295.

upon the experimental design for their safety, it is not clear what activities associated with the peril the participants might focus upon to cause time distortion. Second, asking an observer of a perilous event to try to read some numbers while focusing intently upon activities associated with the peril would certainly interfere with the intense focus that seems instrumental to time distortion.

Concurrent Reporting in Hypnosis As Evidence?

Concurrent reporting in hypnosis is the reporting on an experience, by the subject, as he or she is actually living it, and is common practice in experimental hypnosis.[46] Why not simply have the entranced subject, during hypnosis, tell an investigator what clock times he or she is experiencing? The requirement of a total lack of awareness by the subject of his or her laboratory surroundings, as explained by Dr. Cooper, effectively prevents concurrent reporting. "We avoid [concurrent reporting] because we believe that it tends to prevent the subject from becoming detached from his surroundings in the physical world, and hence from learning time distortion.[47]"

Habeas Corpus of a Sort?

Perhaps there is evidence of time distortion in the bodies of those participants for whom time flowed at a different rate than for other things in their everyday world. Can we collect evidence to demonstrate these aging differences?

Time distortion while in a hypnotic trance probably offers the best opportunity to collect this evidence because it is relatively easy

[46] Cooper and Erickson, 56

[47] Cooper and Erickson, 56

to produce seemingly startling differences in the rate of time flow and because in Dr. Cooper's series of well-documented experiments using hypnosis we have preliminary evidence on which to base further studies. Dr. Cooper has suggested that it might be possible to measure brain activity while a subject is performing a motor activity within the trance state. The "amount of electrical activity per second might be found to vary directly with the amount of hallucinated action, per second of world time."[48] Logic tells us, however, that these efforts would be fruitless. The difficulty is what I have called the "extra body" problem, discussed earlier in this chapter. The hypnotist is in a world created by his or her awareness and that world includes a body, the subject, lying *motionless* on a couch. The body on the couch will display only those symptoms appropriate for any normally-reclining, motionless body. Meanwhile the body of the subject that is performing a motor activity and therefore presumably would be displaying the symptoms expected by Dr. Cooper, is in a different world, a world that does not include the hypnotist and his or her laboratory, and therefore is not accessible to Dr. Cooper to perform his measurements.

Another approach to using the body of the hypnotized subject to provide tangible support for the self-report of time passage may be possible. Notably, there should be age (or aging) differences commensurate with the difference in the rate of time flow. Just as Mick, the stay-at-home twin of the space traveling saga above is able to accomplish more work than his space-traveling brother only because he ages more during their separation, so participants for whom time

---

[48] Cooper and Erickson, 106.

seemed to slow during impending peril and peak experience events must age faster during the event than other observers. Similarly, hypnotic subjects must age faster than the hypnotist. In the world (in the mind) of the hypnotist the body of the subject (lying on his couch) moves through time (a changing space) at the same rate as the hypnotist. In the detached world (in the mind) of the entranced subject, the body of the subject is passing through time (changing space) at a much faster rate than is the hypnotist. This more rapid rate of time passage means that the subject's body is aging much more rapidly than is the body of the hypnotist *and the body lying on the hypnotist's couch*. That (cumulatively very slight, since the rapid aging event occurs for only a short period of time) age difference is impossible to discover until the worlds of the hypnotist and the subject are again shared worlds. In the instant that the subject "wakes" from the trance, the subject must remember (at some level) the hypnotic experience. That the subject can remember the hypnotic experience is necessary (in keeping with the logical imperative) since the subject's autobiographical memory must be complete to fully support the future direction of his or her life—that is, the future direction of the subject's life must be probabilistically related to the entirety of the subject's past life. This, in turn, requires that when the two worlds are again mutually shared, the body of the subject show more aging effects of the experiment than that of the hypnotist. (Subjects usually remembered the hypnotic experience although some remembered only when a pre-trance suggestion had been made by the hypnotist that they remember.[49])

Dr. Cooper's research suggests that this aging effect occurs almost instantaneously, in less than a second of the hypnotist's world

---

[49] Cooper and Erickson, 45.

time after waking the subject since Dr. Cooper "woke" many subjects after only one second or less of world time after they had performed tasks that required more than an hour of trance time. Providentially, science may be very close to a "silver bullet" for measuring the rate of time passage for an individual in a single blood test that can reveal the body's rate of aging at a molecular level.[50] According to the research of scientists at the University of North Carolina, the body's rate of aging at a molecular level is indicated by the concentration level of a protein named "p16IN-K4a," which significantly increases in aging tissue. Blood tests of, for example, the hypnotist and his subject before and after hypnosis, might show differences in the rates of aging over the course of the experiment.

Chapter Summary

Time is an apparent sequential movement at the speed of light through a changing space. This apparent movement occurs because our world is presented to us in frames of time similar to a movie, each frame a separate representation of a universe brought into being by, and coextensive with, our awareness. Each frame is different and each reflects preceding choices made by us within the context of, and shaped by, a preexisting cosmic blueprint.

Since each of us has a separate universe, created within our respective minds, we may each experience somewhat

---

[50] Liu Y, Sanoff HK, Cho H, Burd CE, Torrice C, Ibrahim JG, Thomas NE, Sharpless NE (August 2009). "Expression of p16(INK4a) in peripheral blood T-cells is a biomarker of human aging". *Aging Cell* 8 (4): 439–48. doi:10.1111/j.1474-9726.2009.00489.x. PMC 2752333. PMID 19485966.

different personal times—that is the rates of movement of our bodies through a changing space (movement between the frames of a holographic movie of our lives). In consequence, our bodies also age at somewhat different rates. Under certain circumstances, typically of short duration, the rates at which we move through our changing space, as compared with other things in our everyday world, are markedly different; these include moving at high speeds and altering the scope of our world by selective awareness—by what we choose to observe in that world. Three circumstances of selective awareness that appear to produce marked changes in our personal times are discussed in the chapter: observing or experiencing impending peril, undergoing a peak experience, and participating in hypnosis-assisted time distortion

Questions

Question 1. Recalling that our universe moves through time accompanied by a vast number of other universes, is the holographic film displaying our virtual world showing only a three-dimensional world or does the film really show time progressing for us on a many-dimensional course?

There is evidence at the quantum level to indicate that the latter is true. The Feynman mirror experiment discussed earlier in Chapter One showed that while light is reflected to an observer from all parts of the mirror and though most of the waves are canceled out by other waves before reaching the observer, more than one of the light waves are reaching the observer. *In fact, the mirror has a minimum size below which it no longer serves as a mirror—it will reflect no light to the observer; it must have enough size to accommodate a core of neighboring light waves.*[51] Applying the pointillist principle (that quantum findings may be used to understand macro world events), this experiment suggests that the light source used to display our virtual world, the holographic film of our lives, is displayed in a very large number of dimensions but that we are aware of (light reaches us from) only *a small core of mutually-supportive dimensions.*

Question 2. Continual creation requires that the universe be re-created within each frame of the holographic film displaying our virtual lives. Does that mean that each frame then contains a "big bang" creation event?

John Wheeler felt that his delayed-choice experiment demonstrated that we, by observing the universe, contribute to the *ongoing creation of not just the present and the future but the past as well.*[52] If we consider time to have begun with the big bang (the explosion of dense matter that, according to current cosmological theories, marked the origin of the universe), as does Stephen Hawking,[53] then the answer is "yes"—each frame contains information about a so-called "big bang" creation. We understand this intuitively or we would not be conducting expensive searches within our current frame of time for information about the big bang.

As presented in this chapter, each of us at every moment in time has choices in part determined by our past histories that will in part determine our future histories. These choices and their past

---

[51] Richard P. Feynman, *QED: The Strange Theory of Light and Matter*, (Princeton, NJ: Princeton University Press, 1985, p. 49 and 54.

[52] Folger, interview with John Wheeler.

[53] Stephen Hawking, "The Beginning of Time", http://www.hawking.org.uk/.

and future alternative histories are logically and probabilistically related to our current histories. Our cosmic blueprint unfolds in logical ways (in conformity with the "logical imperative") in response to the choices we make in life given our past and current histories. Critical to this orderly unfolding of our lives is the influence of our past on the ways our futures can unfold. This means that *everything* in our past must be a part of our present—the current frame in the holographic display of our lives.

While it may seem that conditions in our distant past could have no importance to our current lives, this is not so. Life is possible in our universe only because certain conditions in our universe are exactly as they are; if these conditions were just a little off, life would never have evolved.[54]

> The fine-tuning of seemingly heterogeneous values and ratios necessary to get from the big bang to life as we know it involves intricate coordination over vast differences in scale—from the galactic level down to the subatomic one— and across multibillion-year tracts of time.[55]

The past relative to our present includes everything that has happened to our universe from the instant of its creation.

The mechanism bringing the past into the present might be the holographic nature of the universe. If the universe is holographic, as it is believed by many scientists to be, it conforms to the holographic principle which states that, for any given area of space,

---

[54] For a discussion of the anthropic principle, see Martin Rees, *Just Six Numbers: The Deep Forces that Shape the Universe*, (New York, NY: Basic Books, 2000), pp. 6-9.

[55] Glynn, Patrick, *God: The Evidence* (Rocklin, CA: Prima Publishing, 1997), p. 19.

all the information contained within the space is also present on the surface of the space.[56] Maldacena cautions, however, that an expanding universe, such as ours, that comes from a big bang, does not have a well-behaved boundary. Consequently, it is not clear how to define a holographic theory for our universe; there is no convenient place to put the hologram.[57] Perhaps, however, those surfaces of space that are snapshots of the universe at one instant in time—the frames of our holographic movies, if you will—might provide a well-behaved boundary for the hologram.

Holographic theory aside, if, from the beginning of time, each frame has contained all the information contained in all frames to that point in time, no information about our lives is ever lost. As new frames occur, each new frame is supported by—coherently contains—all information in all previous frames. Thus information about our universe is preserved from its beginning. Each frame of the holographic film displaying our virtual lives contains a "big bang" creation event.

---

[56] See "Leonard Susskind, Bad Boy of Physics," interview by Peter Byrne, *Scientific American*, July 2011, 80-83; Jacob D. Beckenstein, "Information in the Holographic Universe," *Scientific American*, July 14, 2003; and Juan Maldacena, "The Illusion of Gravity," *Scientific American*, November 2005, 56-63.

[57] Maldacena, 58, 63. Maldacena cautions, however, that the mathematics of the theory has not yet been rigorously proved for ordinary space.

CHAPTER 3
TIME AND PARTICLES IN SUPERPOSITION

> Latent structure is master of obvious structure.
> (Heraclitus)[1]

The model of time presented in Chapter 2 helps us to under-stand one of the most problematic concepts of physics: particles in superposition. Particles are said to be in superposition when they exist in all their theoretically possible states as determined

---

[1] Charles H. Kahn, *The Art and Thought of Heraclitus: A New Arrangement and Translation of the Fragments* (Cambridge, UK: Cambridge University Press, 1979), 35.

by the wave function of the particle. Two concepts critical to understanding particle superposition are Bell's theorem and the many-worlds interpretation of reality (MWI). At the time of this writing, we are unable to explain fully an important aspect of Bell's theorem. And MWI presents a problem in that many physicists do not accept it as a correct picture of reality because they do not feel there is sufficient evidence for its veracity.[2] Both Bell's theorem and MWI can be better understood by viewing them in terms of the holographic movie model of time (HMM) presented in Chapter 2. A better understanding of each can, in turn, open the way to an understanding and a more efficient use of the quantum properties of elementary particles in superposition and perhaps to important advances in knowledge of our everyday world. I will discuss these points in this chapter.

The Holographic Movie Model of Time and Bell's Theorem

Bell's theorem is a mathematical construct which, as such, is indecipherable to the nonmathematician. One of the implications of Bell's theorem, however, is that, at a deep and fundamental level, the spatially-separated parts of the universe are connected in an intimate and immediate way. Thus reality in one location, which as we know is brought into being by measurement, appears to be influenced by

---

[2] Max Tegmark, "The Interpretation of Quantum Mechanics: Many Worlds or Many Words?" Institute for Advanced Study, Princeton, NJ 08540; max@ias.edu, (September 15, 1997), 1; David Deutsch, "Three Connections between Everett's Interpretation and Experiment," in R. Penrose and C. J. Isham, *Quantum Concepts in Space and Time* (Oxford, Eng.: Clarendon Press, 1986), 215-225.

a measurement performed *simultaneously* at a distant location. If, for example, as shown in the so-called Aspect experiment,[3] you measure, say, the spin of one particle, this appears to affect the likelihood of obtaining *simultaneously* an analogous measurement of the spin of a so-called "entangled" particle (a particle that has previously interacted physically with the first particle so that their wave functions are correlated) in another distant location. This result Einstein termed "spooky action at a distance."[4]

The resolution of this apparent quandary, in terms of the holographic movie model of time presented in Chapter 2, is this: The entire Aspect experiment, which demonstrated what appeared to be action at a distance, *in view of the simultaneity of the measurements*, occurred not *across* holographic movie time frames but *within* a single time frame. No change occurred *during the experiment* since, as explained in the discussion of real time and comparative time, and as illustrated in the twin paradox saga, *change occurs only between time frames*. And if no change has occurred, no action has occurred (including action at a distance). At the time of the measurements, each of the entangled electrons *already was in the state of spin it would be found to be in when measured*. Each "knew" which state it was in; only we did not know. What had appeared to be

3 See Alain Aspect and Philippe Grangier, "Experiments on Einstein-Podolsky-Rosen-type Correlations with Pairs of Visible Protons," in R. Penrose and C. J. Isham, *Quantum Concepts in Space and Time* (Oxford, Eng.: Clarendon Press, 1986), 1-15.

4 Charles Seife, *Decoding the Universe: How the New Science of Information is Explaining Everything in the Cosmos, from Our Brains to Black Holes* (New York, NY: Penguin Group (USA) Inc., 2006), 178-179

separate measurements of two particles was actually two measurements of the same particle within a single human time frame. The question remains, however: *While one measurement was made by scientists conducting the Aspect experiment, who* made the other measurement and *where was he or she located?* We will return to this question below.

The Holographic Movie Model of Time and the Many-Worlds Interpretation

As stated above, some physicists do not accept the many-worlds interpretation (MWI) as a correct picture of reality. The many-worlds interpretation, discussed above, views reality as a many-branched tree, wherein every logically possible outcome is realized and exists in its own separate world. If there are two logically possible outcomes, then change will produce two new branches of the tree, each a different outcome; and the outcomes will be perfectly anticorrelated—if we know one, we know the other.

As shown in the Aspect experiment cited above, the holographic movie model of time (HMM), *in view of the simultaneity of the measurement*, requires that the two mirror images of a particle in superposition all exist in one human time frame and, since change occurs only between time frames, no change can have occurred to the particles. A general principle may be stated: superpositions of particles all appear at the same instant of time in our world. We know this not only because of what HMM tells us but also because all except the one particle we choose to measure "disappear" (insofar as we are concerned) at the same instant with wave function collapse. We know that only one particle is located in our world for when we measure a particle in superposition we find only one particle. A key question, then, as stated above, in connection with Bell's theorem, is this: *Where are the "extra" particles located?* The many-worlds interpretation would suggest that they are in other worlds of the multiverse.

Experience with quantum computers also provides evidence that supports MWI. (While no large-scale quantum computers have yet been built, "proof of concept" quantum computers have been developed.[5]) Quantum computers differ from digital computers in that while digital computers encode data using binary digits (bits), quantum computers use qubits (quantum properties of elementary particles such as spin, in superposition of all their possible states) to represent data. A single qubit can represent a one, a zero, or, crucially, any quantum superposition of these two qubit states; a pair of qubits can be in any quantum superposition of four states, and three qubits in any superposition of eight states. In general, a quantum computer with η qubits can be in an arbitrary superposition of up to 2^η different states *simultaneously*; this contrasts with a normal computer, which can only be in one of these states at any one time. Quantum computers thus can hold and use exponentially more information than classical computers, yet when the final state of the qubits is measured, they will only be found in one of the possible states they were in before measurement.[6] Where was this large amount of information in these configurations of possible states before it was lost in the measurement process? We know it is not in our world for when we look for configurations, we invariably

---

[5] "Quantum Computer," Wikipedia, February 21, 2013.

[6] See George Johnson, *A Shortcut through Time: The Path to the Quantum Computer* (New York, NY: Alfred A. Knopf, 2003), for an excellent, though challenging, description of quantum computing.

find only one, the one associated with the qubit in our world. The MWI answer would be: the unaccounted-for information was in the many worlds of the multiverse.

David Deutsch, an ardent supporter of MWI, adopting the viewpoint of a future computer programmer using a quantum computer, asks an even more challenging question. The task of the computer programmer using a quantum computer is to establish, using quantum logic gates, a sequence for manipulating the qubits that will not prematurely collapse a superposition before measurement of all the desired states has been accomplished. Deutsch asks:

> What will a practical quantum programmer make of the statement ... that his description of his program as proceeding in a sequence of parallel steps is 'meaningless', that only the final result exists physically? He will know from everyday experience that the execution of the steps is real in every way that counts to him: all his ingenuity is devoted to planning precisely which operations are to be executed in which order and in which Everett branch. To use N-fold quantum parallelism he must write N independent programs. And if he makes a mistake in his coding for any one of them, the whole program will not work. If told that the Everett branches do not exist in reality, he might well ask—and I issue this as a challenge to Everett's opponents—'On the occasions when a quantum processor delivers the results of N processor-days of computation within a day of being programmed, *where was the computation done?*[7] (Emphasis Deutsch)

---

[7] See David Deutsch, "Three Connections between Everett's Interpretation and Experiment," in Penrose and Isham, *Quantum Concepts in Space and Time* (Oxford, Eng.: Clarendon Press, 1986), 219–220.

If and until a better working hypothesis is formulated of where these "mirror image" particles are located and where the computations of quantum computers are carried out, the most reasonable answer is: they are located in the many worlds of MWI.

Future Research about Particles in Superposition

A better understanding of particles in superposition is made possible by viewing these topics through the lens of the holographic movie model of time, and a better understanding of what particles in superposition are may be especially important for advances in knowledge. Two areas of research seem especially likely to benefit from a better understanding of what particles in superposition are: research about how we think creatively and research to better understand how we may use particles in superposition as basic units of action in a growing number of practical applications.

How We Think Creatively

With better understanding of superposition, we may begin to better understand what we do when we think, and especially, perhaps, when we think creatively. For example, it now seems clear that the information discovered by creative thinkers and performers, discussed in Chapter 1 above, likely is in the minds of their many-worlds counterparts. That is, creative persons are able to hold many contrasting ideas of their many-worlds counterparts in superposition until some event causes collapse of the wave function and the emergence of one final configuration of ideas.

In this context, consider a type of thinker that we had not previously discussed that is quite interesting in this regard—a person with savant syndrome who often has a very low IQ but displays brilliance in a specific area, especially one involving memory, such as rapid calculation, art, or musical ability.[8] A study by L. Pring and B. Hermelin of an autistic savant confirmed findings of other studies of autistic savants in that the savant's exceptional abilities in a specific area appeared to be attributable to his or her cognitive style:

> … a cognitive style of 'weak central coherence' as adopted
> by autistic savants may protect single representations from
> being retained in the form of stable enduring wholes, and
> … such a segmentation strategy may allow for the transfor-
> mation, reorganization and reconstruction of the relation-
> ship between single items of information.

Remembering that the key to successful programming of the quantum computer is to establish with our computer programs the terms on which wave-function collapse occurs so that the vast amount of information in the configurations of possible states is not lost in the measurement process, is the above description of the thought patterns of an autistic savant not an example of quantum computing by a human mind? If so, what could the autistic savant tell or learn from a quantum computer programmer?

Employing the Particle in Superposition as a Basic Unit of Action

A better understanding of what particles in superposition are also should serve to advance knowledge in a growing number of

---

[8] L. Pring and B. Hermelin, "Numbers and Letters: Exploring an Autistic Savant's Unpracticed Ability." *Neurocase*, 2002, 8(4), 330-337.

fields of study that employ the particle in superposition as a *basic unit of action*, fields as diverse as quantum computing and condensed matter. Subir Sachdev, a condensed-matter physicist who studies materials such as metals and superconductors, and others in his field, have discovered, in Sachdev's words, "our materials doing things we never thought they could. They form distinctively quantum phases of matter, the structure of which involves …. what we call quantum entanglement." In a typical example of material in the laboratory, according to Sachdev, 10^{23} electrons in superposition may be involved in an unimaginably complex web of quantum entanglement among the electrons—so complex that it is beyond the ability of physicists to describe directly.[9]

Are trade-offs possible among the above areas? Even if trading of ideas among these fields of endeavor is not possible, it is difficult to imagine areas of study with as much promise as particles in superposition and their residence—the many worlds of Hugh Everett. I might say (with apologies to my readers) the skies of the many worlds are the limit!

---

[9] Subir Sachdev, "Quantum Physics: Strange and Stringy." *Scientific American*, January 2013, 44-51.

Chapter Summary

From an engineering standpoint, we understand elementary particles, as such, quite well—that is we can use our understanding of them for important applications in everyday life.[10] We do not, at this point, however, understand well particles in superposition. Particles are in superposition when they exist in all their theoretically possible states as determined by the wave function of the particle. Two concepts critical to understanding particle superposition are Bell's theorem and the many-worlds interpretation of reality (MWI). Both Bell's theorem and MWI can be better understood by viewing them in terms of the holographic movie model of time presented in Chapter 2. A better understanding of each can, in turn, open the way to an understanding of, and a more efficient use of, the quantum properties of elementary particles in superposition and perhaps to important advances in knowledge of our everyday world.

---

[10] See Fritjof Capra, *The Tao of Physics: An Exploration of the Parallels between Modern Physics and Eastern Mysticism* (Berkeley, CA: Shambhala Publications Inc., 1975), pp. 51-64.

CHAPTER 4
INERTIA, GRAVITY, AND TIME

> We are trapped by time in an incomplete world. The designer of this universe mercifully realized our incompleteness and gave us gravity to hold us together and allow us to exist for a moment of time as incomplete wholes. (Donald W. Jarrell)[1]

In this chapter, as the title suggests, we will discuss how inertia and gravity should be understood in the framework of the holographic movie model of time. Inertia, as the reader will know, is a

---

[1] One of the author's own writings entered December 3, 1988, in an unpublished private collection of favorite poems and quotations.

property of matter by which it continues in its existing state of rest or uniform motion in a straight line, unless that state is changed by an external force. In our daily lives, inertia is the force we feel when we change the rate (or direction) of movement of an object having mass. Gravity is the natural force of attraction exerted by any body having mass, such as our planet Earth, upon objects at or near its surface, tending to draw them toward the center of the body. As suggested by the fact that mass is a key component of the definition of both inertia and gravity, the two forces are inextricably linked—in fact, when both are present (as when an object is in free fall), they are numerically equivalent. As indicated by the title for this chapter, the holographic movie model of time, as presented in Chapter 2, allows us to form a reasonable hypothesis as to what causes the inertial and gravitational forces and how these forces alter comparative times of all of us.

The Linkage of Inertia and Time

We will look first at the linkage of inertia and time but before I can do this, I need to provide a background for the answer from the material of Chapters 1 and 2.

As discussed in Chapter 1 above, each of us has a virtual, personal, world and each of us has a projector (or equivalent mechanism) that creates this personal world. Where is that personal projector? We know from the presentation in Chapter 1 that, in the dual-realities framework, neither the projector nor the virtual images of our scripted reality are "out there somewhere"; the images of our everyday world, our multiverse, cannot be projected to us by those same images. The script of our everyday lives, our universe, must therefore come to us from within, and the projector itself must be located deep within us, beyond the multiverse, beyond space and time. Where then is the projector in relation to our virtual body. Again, as indicated in Chapter 2, logic (and the dual-realities view of reality) tells us that the images of our everyday world must be projected to us from all directions because if you travel in any direction, time slows since you are moving toward your projector. (Note that you cannot move away from the projector, since it is everywhere; therefore, your movement relative to other objects always has a slowing effect on your comparative time.) Summing up the Chapters 1 and 2 review presented here, we each have a quantum-level-distributed projector transmitting from an infinite number of projection points a virtual image of a universe at one point in time.

What, then, is inertia? As indicated in Chapter 2, we are moving through a changing space (moving between the frames of a holographic movie) at the speed of light. Why the speed of light? Because light (electromagnetic force associated with the exchange of virtual photons) is the medium used to transmit the holographic movies of our lives. And how do we know that we are moving at the speed of light? First, we know we are moving at the speed of light because the speed of light is invariant for us and is the maximum speed any object may travel in our multiverse—moving at a speed exceeding that of light is impossible since our world is being created at the speed of light. Second, we know we are moving at the speed of light because, in the framework of the holographic movie model of time, comparative time slows for a moving object and "time stands still for an object moving at the speed of light."

We are normally unaware of this movement through space at the speed of light because the rate of movement is unchanging and because, since everything in our awareable universe is moving with us, we have no point of reference against which to notice this movement. You may have experienced a somewhat similar phenomenon in an enclosed elevator that is moving at a uniform rate. You have no inertial effect allowing you to detect the uniform movement but when the elevator slows or accelerates, you feel an inertial force telling you that you are (or were) moving. In a similar manner, when we move relative to the rest of our world and thus slow the rate at which we move between frames of the movie, we produce an inertial force. Movement through space, by slowing movement through time produces an inertial force.

We can now turn our attention to an explanation of the gravitational force.

The Linkage of Inertia, Gravity, and Time

As stated earlier, in the framework of the holographic movie model of time, we each have a quantum-level-distributed projector transmitting from an infinite number of projection points a virtual image of a universe at one point in time. On its face this sounds like a potentially chaotic situation, a recipe for disaster. But, as discussed in Chapter 2, our universe is not chaotic but is rationally ordered in consistency with a cosmic blueprint; given this potentially chaotic situation, how does our universe achieve the exquisite order it displays?

I suggest that each projection point transmitting our virtual world at any given time is represented by a uniquely identified fermion—those elementary particles usually associated with matter (as contrasted with bosons, which are generally force-carrier particles). This unique identification is accomplished by assigning each particle a unique time stamp: fermions are created one at a time, giving each a unique place (order) in time. (In a similar manner, you can think of computer memory as an array of storage boxes, each of which is one byte in length; each "box" has an address, a unique number, assigned to it.) That fermions may indeed be uniquely identified is suggested by the Pauli exclusion principle, which states that only one fermion can occupy a particular quantum state at any given time. If each fermion is uniquely identified, this provides a code for use in the cosmic blueprint (perhaps written as a set of executable instructions) to guide the composition of our everyday world.

One of the details that would be required of a cosmic blueprint of our everyday world would be the measured "amount" of matter that physical objects contain. (Please bear in mind that, as discussed in Chapter 1, elementary particles are not actual physical "things" to be "contained" but are abstract constructions and have no reality beyond the measurement records we compile.) Massive bodies such as the earth, by definition, are those that are measured to contain a large quantity of matter (fermions) confined to a relatively small space. Inasmuch as time is defined by reference to spatial distance (for example, your image in the mirror is only a tiny fraction of a second behind you in time while a distant star may be thousands of light years behind you in time), nearness in space means that all of these uniquely identified particles in a massive body will need to be presented in the script of our everyday lives at nearly the same time; this requirement likely causes a modification in the way the projector of the script of our daily lives displays these massive bodies. If fermions are projected one at a time (mimicking the operations of most modern computers, which carry out one instruction at a time but do that simple task in an incredibly short period of time), it follows that a massive body, which contains a large quantity of fermions confined to a relatively small space, would take longer to project than would a less massive body of the same size. The quantum-level-distributed projector from which the script for our everyday world emerges likely is slowed by this requirement for the projection of many elementary particles at almost the same time just as (drawing on another computer analogy) a printer is slowed when the amount of detail to be resolved on a page of copy is increased. When the projection of a massive body is slowed, passage through time of the massive body also is slowed. But just as the printing of a detailed image on one page by a printer slows not only the printing of the image itself but of the entire page, the projection of a massive body will slow the projection of everything in the vicinity of the massive body that follows it in the projection sequence, everything in its "gravitational field"—in theory, everything in the universe but of decreasing importance as distance from the center of the massive body increases. This means that if you are in close proximity to the center of gravity of a massive body such as the earth, which is moving slowly between frames of the holographic movie of your life, the rate at which you move between frames of the holographic movie of your life also will be slowed as compared to that of objects farther from the massive body. Your movement through time (and the comparative rate at which you age) will be slowed so long as you continue to be near the center of gravity of the massive object.[2]   Just as movement through space caused a slowing of movement through time which, in turn, caused an inertial force, so too, a slowing of movement through time caused by proximity to a massive body will produce an inertial force which, by convention, we call the gravitational force.

The time dilation effect of gravity may also appear at the quantum level. Arguing backward from the pointillist principle, which states that findings made through observations at the quantum level can be applied at the macro level, the principle suggests that a time dilation effect in the external world has its counterpart in the quantum world. Albrecht Giese has suggested that the counterpart to gravitational effects in the macro world emerges at the quantum level in the kinematic processes of particles. These kinematic processes produce a force tending to compel the movement of objects at or near the surface of a massive body to move toward its center. Giese's model has been criticized in that it seems to assume an absolute rest frame. If, however, the

---

[2] A topographical representation of the rate of passage of comparative time for any given area of space containing a massive body thus will display the Einsteinian curvatures of space.

model of time presented in Chapter 2 above is correct, there is in fact an absolute time and therefore an absolute rest frame.

Free Fall and Numerical Equivalence

The special case of free fall helps to illustrate the previous discussion in this chapter. Free fall is the term used to describe movement under the influence of gravity only. Inertia and gravity are numerically equal for an object in free fall toward a massive object such as the Earth—what mechanism brings about this exact equality? The holographic movie model of time may provide an answer.

Given the previous discussion, we know that the object in free fall near (by convention "above") the Earth is simultaneously subject to two separate influences. Comparative time is being slowed for the object because it is near a massive object (the Earth). The closer the falling object comes to the massive object, the more slowly comparative time passes and the greater the gravitational force. Comparative time also is being slowed for the object because of its movement through space toward Earth. The faster the falling object falls, the more slowly comparative time passes for the object and the greater the inertial force. The gravitational and inertial forces, are not the cause of a slowing of comparative time for the falling object but are themselves the result of a slowing of movement through time. As such they must be numerically equal. Here is why.

As the object approaches the center of gravity of the Earth, the slowing effect of the Earth's mass upon its comparative time (and the gravitational pull) on the falling object increases and it moves more rapidly. As the object begins to move more rapidly, slowing of its comparative time (and the inertial force) also intensifies. When the falling object is, let's say, 500 feet above the earth, it has a unique place in the spectrum of time—as explained above, the object consists of a collection of fermions each with a unique time stamp. There cannot be two different times associated with an object whether in free fall or otherwise. The two forces associated with comparative time for the object must, therefore, be numerically equal.

Chapter Summary

This chapter discusses how inertia and gravity should be understood in the framework of the holographic movie model of time.

Inertia is the force we experience when we change the rate or direction of movement of an object having mass since this has the effect of slowing the rate at which the object moves through a changing space (moves between the frames of the holographic movie of our lives). Gravity is the force we experience when nearby massive bodies, such as the Earth, are slowed in their passage through time as a result of the slowing effect of these bodies on the projection of the movies of their, and our, daily lives.

Free fall of an object (movement under the influence of gravity only) is used to explain the numerical equivalence of inertia and gravity and to show that the holographic movie model of time can provide practical explanations of what gravity and inertia are at a fundamental level.

CHAPTER 5
LIFE AND MEANING IN A VIRTUAL WORLD

Losing an illusion makes you wiser than finding a
truth. (Ludwig Börne)[1]

---

[1] While clearly this quotation is attributable to Börne, it has proven impossible
to find an English-language citation. For a quite-interesting discussion of the
difficulties involved (and of the context of this quote), see quoteinvestigator.
com/2013/04/04/losing-illusion.

> Man did not weave the web of life. He is merely a strand in it. Whatever he does to the web, he does to himself. (Anonymous)[2]

> He has told you, O mortal, what is good; and what does the Lord require of you but to do justice, and to love kindness, and to walk humbly with your God? (Micah 6:8)[3]

Previous chapters have presented what I believe are challenging views of reality and time. If accepted as correct, these new ways to view reality and time suggest *a much more important role for the individual self* in relation to all else than in previous views of reality and time. This shift in the relative importance of the individual self requires, at a minimum, a new look at the following questions about individual life and meaning:

What is life?

What is your extended universe?

Do our lives have purpose?

If our lives have purpose, what is that purpose?

Is there a Creator and/or a creative process?

---

[2] This popular quote is, by tradition, attributed to Chief Seattle, leader of Native American tribes, as part of a speech made in 1854 There is little evidence to supprt this belief. The words are believed most likely to have been written by an unknown screen writer in 1971. See "Chief Seattle." *Snopes*. Sept. 26, 2007.

[3] Micah 6:8.

I will preliminarily examine these questions in the pages below.

What Is Life and the Universe?

Chapter 1 suggests that life is an *awareness* that brings into being a multiverse that consists of our own universe of collapsed wave functions together with innumerable counterpart universes that exist *for us* as uncollapsed and unseen waves. The outward manifestation of being alive is the existence of a virtual body and its extended multiverse. That we are not aware of most of the multiverse accounts for the failure of wave functions of other worlds to collapse. Logic dictates that we cannot collapse these wave functions since the unique persons we are cannot be in more than one world. While we may have counterparts in other worlds, "we" cannot be there. (But, as discussed in Chapter 3, quantum computer programmers *can* retrieve information from other-universe databases without collapsing other-universe wave functions and creative thinkers can do the same with their own version of quantum computing.)

This definition would suggest, then, that life and time in our multiverses are coexistent with our own lives; each begins when our virtual worlds begin and ends when our virtual worlds end. What is our portion of the multiverse, our universe?

The universe, in the dual-realities view, is the stage upon which our everyday lives take place. This raises a question: Why are our stages so vast, appearing to extend in both time and space far beyond what would, at a first thought, be necessary to accommodate our virtual reality lives. But this first thought is incorrect. The confusion arises, I believe, because we greatly underestimate the complexity of human life, perhaps because we tend to think of this human life as consisting in its entirety of a very complex robotic body. It is difficult to overstate the magnitude of this error, but the nature of the error is expressed in part by remembering that the human body is organic while robots are made of metal or other inorganic material. Further, only the conscious human mind (no robot or other mechanical device) is believed to be able to cause the collapse of wave functions that brings into being the vast universe that serves as the stage for the consciousness that brought it into being.[4]

Why, then, is this stage upon which our everyday lives take place, our universe, so vast? Our logic, in the form of scientific studies, tells us that the organic human being, with that important trait of being conscious, is the product of an almost incredibly time-and-space-consuming evolutionary process within a universe remarkably fine-tuned to allow this process and life to happen. If we assume that the evolutionary process that produced this universe occurred, to this point, without significant intervention then, just as we assume that the 8-year old oak tree growing in a dense forest required 8 years in that location to reach its current size, we also must assume that this vast stage was required to support the evolutionary process that brought you into being. You as an individual person are fully defined as an individual only by the evolutionary process that produced you and your nurturing universe.

---

[4] See Tim Folger, interview with John Wheeler, "Does the Universe Exist if We're Not Looking?" *Discover*, June 1, 2002. See also the discussion of the "delayed-choice" experiment, in the text above.

The obvious question that now arises is, does this marvelous creation, "you," have a creator? Or did it simply emerge, perhaps by chance, from some inexplicable creative process?

Is There a Creator?

There is an unavoidable dilemma we face when we try to answer this question, "Is there a Creator?" We can say nothing (literally) about the *nature* of a creator of our universe, including whether a creator has existence. Logic tells us there is not a tangible being that created us that we might call either "God" or "Creator." Any "thing" that created our universe would have to exist beyond time and space where no things or time-contained actions (events) can exist and existence itself has no meaning. (So even the word "thing" that I just used is not allowed in logical discourse.)

And yet, logic works faultlessly *within the confines of my existence* and that logic tells me that our (virtual) multiverse almost certainly comes to us from outside the multiverse itself—the images of our many worlds cannot be projected to us by those same images. To project from inside our virtual multiverse in full detail that same multiverse would seemingly create an energy shortage: there would be no surplus energy with which to accomplish the projection. The script of our everyday lives, our universe, must therefore come to us from beyond space and time.

Further, if my Chapter 2 representation of time is correct, and *creation is a continuing process*, the creation *continues* to be sustained from beyond space and time. While I have avoided saying that there is a creator, most persons would likely answer "yes" to the following question: "Isn't it logical to assume that the continuing presence of a creative force beyond our universe suggests anew that there is a creator?"

And yet, the simultaneity of irreconcilable ideas remains and is not unexpected; it is the everyday equivalent of the singularity of mathematics that effectively prevents a mathematical description of the creation process.[5] The proper term to use for the creative event that produced life, then, is the word "miracle" in the sense of "something not explicable by natural or scientific laws."

While we cannot answer the question, "Is there a creator," the question of whether our lives appear to have purpose remains. I will now consider this question.

Do Our Lives Have Purpose?

Scientists are understandably reluctant to attribute purpose to life since this sounds like the anthropocentric error of trying to explain everything from a human viewpoint. However, with increased understanding of our universe, it has become increasingly difficult to deny that our lives have purpose.

The primary argument used in the past to support the contention that human life is purposeful has been the anthropic principle, a principle that intrigues physicists and philosophers alike. Why do the values of the universe's fundamental constants happen, in

---

[5] For a discussion of the "*initial* space-time singularity which now represents the big bang, in any (appropriately) expanding universe," see Penrose, *The Emperor's New Mind: Concerning Computers, Minds, and the Laws of Physics* (New York, NY: Oxford University Press, 1989), 336-338.

spite of incredible odds to the contrary, to fall within a narrow range thought to be compatible with life? To many, this suggests that perhaps the universe exists for the express purpose of supporting (human) life. Others have argued that only in a universe capable of supporting life will there be living beings capable of observing any such fine-tuning, while all universes less compatible with life will go unbeheld; purpose is therefore not needed to explain why life exists.[6]

If, as posited in Chapter 1, we each have a personal universe (and accompanying multiverse), this fact would provide powerful support for the belief that the universe exists for the express purpose of supporting (human) life. And the anthropic argument is turned on its head. If our bodies and personal universes are manifestations of our being alive, then of course the universe of our (intangible) everyday life would be suitable for human life; it is the expected outcome—why should we expect a different result? And the incredible fine-tuning of the universe required to support human life together with the marvelously creative evolutionary process certainly suggests a concern by something to get the creation right, a concern perhaps driven by a purpose? Further, and perhaps most compelling, if life comes into existence in the context of a cosmic blueprint that shapes our lives, it strains credulity beyond belief that life can be without purpose. Here, of course, I am using human logic. But I am, after all, a human being subservient to the "logical imperative." And in terms

---

[6] See Davies, *The Cosmic Blueprint*, 160-163. Also see Penrose, *The Emperor's New Mind*, 405-406, and P. C. W. Davies, "Where Do the Laws of Physics Come From?" www.scribd.com/doc/.../ Where-Do-the-Laws-of-Physics-Come-From, 9.

of that logic, it has become quite difficult indeed to explain away the elephant in the room called "purposeful life."

And what is that life purpose?

Our Primary Purpose in Life: to Choose between Good and Evil

I believe the overriding purpose of life is to choose between good and evil. (I do not think that this is the only purpose of life. I think, for example, that a quite strong case could be made that another purpose of life is simply to enjoy it.) While I cannot offer *proof* of this contention, I can offer the following arguments to support this view: major world religions support the view; people are by nature morally evaluative; this view resolves a long-standing philosophical problem called the "problem of evil;" and this purpose offers an explanation of why our virtual worlds contains other life forms.

The Major World Religions Support This View

All major world religions, while they have different understandings of what good and evil are, urge their followers to do good and refrain from doing evil.[7] In fact, the Hebrew Bible, sacred text of the Abrahamic faiths—consisting of Judaism, Christianity, Islam, and the Baha'i Faith and accounting for 54 percent of all the world's religious adherents—suggests that not only are people to choose between good and evil but that good and evil *were brought into the world to allow this choice.* In Deuteronomy 30:15-19, God, speaking

---

[7] Chapter 9: "Good and Evil," in Philip H. Phenix, *Intelligible Religion*, religion-online.org

through Moses (considered a prophet or spokesperson of God by all the Abrahamic faiths) to his chosen people, the Israelites, says:

> See, I have set before you today life and good, death and evil. If you obey the commandments of the Lord your God that I command you today, ... then you shall live and multiply..... But if your heart turns away, and you will not hear, ... I declare to you today, that you shall surely perish. .... I call heaven and earth to witness against you today, that I have set before you life and death, blessing and curse. Therefore choose life, that you and your offspring may live ....[8]

In God's plan as set forth here, God was responsible for bringing both good and evil into the world with the apparent purpose of giving to his chosen people and, by extension, in the belief system of the later-emerging Abrahamic faiths, to all humankind, a choice. Humans were not to be automatons who could only do God's bidding but were to be allowed to choose between good and evil—they were to be masters of their own destiny.

Other Biblical texts also explicitly state that the God of the Israelites is directly responsible for bringing evil into the world to test his people. In Genesis 18, God, because of the importance of Abraham in God's plans for the Israelites, reveals to Abraham that God will destroy the entire cities of Sodom and Gomorrah, both the good and the evil people, because of the sins of the many; Abraham apparently passes the test by pointing out that this is not a just action and therefore not worthy of the judge of all the

---

[8] Deuteronomy 30:15-19, English Standard Version.

Earth bringing God to retreat from his original plan.[9] In Exodus, Chapters 4-15, God is said to have hardened Pharaoh's heart so that it will be necessary to bring plagues against the Egyptians to convince Pharaoh to let the Israelites leave Egypt, this done so that the Egyptians and the Israelites will know that the Lord is God.[10] And in Job, Chapter 1, God is said to have allowed Satan to destroy innocent members of Job's family as a test of Job's love for God.[11]

What credence should be given to the testimony of these writers of the Christian Old Testament and the Hebrew Bible? Modernists may undervalue this literature. The writings themselves testify to the extensive examination by the writers of the human condition (the unique and inescapable features of being human) and of the ways in which the Israelites made their choices between good and evil. And these choices were made within a continuing relationship with a God who warned them that their choices would determine the quality of their lives as a people here on Earth. These writings, then, are the testimony of very motivated and astute observers of the human condition. An objective reading of the history of the Israelites in the Hebrew Bible will convince most people that this *is the story of humankind*; substitute names and places and it is the story of humankind today and throughout history.

We Are by Nature Morally Evaluative

But is this the way we actually behave? Are our lives really concerned in an important way with choices between good and evil?

---

[9] Genesis 18: 16-33, English Standard Version.

[10] Exodus 4-15, English Standard Version.

[11] Job 1, English Standard Version.

Two streams of scientific thought suggest that they are; to an astonishing extent, we distinguish between good and evil in our daily lives.

Attitude studies across the world in a wide range of cultures using the semantic differential technique developed by Osgood, Suci, and Tannenbaum show that three factors (derived from factor analyses of large collections of semantic differential scales) largely determine the connotative meaning of things in the daily lives of people.[12] Connotative meaning has to do with, not the literal meaning of a thing, but the feelings something arouses when we think of it. These factors are: evaluation, potency, and activity. Evaluation loads highest on the scalar adjective pair 'good-bad', the 'strong-weak' adjective pair defines the potency factor, and the adjective pair 'active-passive' defines the activity factor. The evaluative factor ordinarily accounted for the largest amount of variability among scale ratings and was identified by Osgood and his colleagues as synonymous with attitude (the tendency to act).[13]

Independent of the above studies, neuroscientists using neuroimaging (magnetic resonance imaging) have found that making decisions about right and wrong are instinctive for modern humans. Unlike nonmoral or impersonal dilemmas, where we make logical

---

[12] C.E. Osgood, G. Suci, and P. Tannenbaum, *The Measurement of Meaning* (Urbana, IL: University of Illinois Press).

[13] See S. Himmelfarb, "The measurement of attitudes," in *Psychology of Attitudes*, A.H. Eagly and S. Chaiken (eds.) (New York, NY: Harcourt Brace Jovanovich, 1993), 23-87, and David R. Heise, *Surveying Cultures: Discovering Shared Conceptions and Sentiments* (Hoboken NJ: Wiley, 2010).

decisions, personal moral dilemmas such as the following trigger instinctual emotions that strongly influence the way we respond to the need to act:

> You are running down a crowded corridor in the airport, trying to catch a flight that's about to leave. Suddenly, an old woman in front of you slips and falls hard. Do you stop to help, knowing that you'll miss your plane?[14]

Dilemmas such as this arouse emotions that play a powerful role in our moral judgments, triggering instinctual responses. These instinctual responses, neuroscientists believe, are the product of millions of years of evolution. Evolutionary biologist and cognitive neuroscientsz Marc Hauser argues that natural selection has molded a universal moral grammar within our brains that enables us to make rapid decisions about ethical dilemmas. "There appears to be some kind of unconscious process driving moral judgments without its being accessible to conscious reflection."[15]

---

[14] See Carl Zimmer and Jeff West, "Whose Life Would You Save?: Scientists Say Morality May Be Hardwired into Our Brains by Evolution," interview with philosopher and neuroethicist Joshua Greene, *Discover*, April 21, 2004, 2. For a report showing quite similar results using more conventional study techniques, see Adrian F. Ward, "Scientists Probe Human Nature--and Discover We Are Good, After All; Recent Studies Find our First Impulses Are Selfless, *Scientific American*, November 20, 2012, 23.

[15] See Josie Glausiusz, "The Discover Interview: Marc Hauser," *Discover*, May 2007, 62-66. See also Marc Hauser, *Moral Minds: How Nature designed our Universal Sense of Right and Wrong* (New York, NY: Harper Collins, 2006)

The answer to the above question, then, is "yes." We are concerned in our daily lives to a very significant degree with choices between good and evil.

Solving the "Problem of Evil"

A third reason I suggest that the purpose of life is to choose between good and evil is that it addresses the "problem of evil," which is this: although virtually all world religions feel that their god is both good and all-powerful, evil appears to be in plentiful supply in the world. As formally defined in the philosophy of religion, the "problem of evil" is framed as the question of how to reconcile the existence of evil with belief in a deity who is simultaneously omnipotent, omniscient and omnibenevolent. [16] But if the evil that we see here is intended to allow a choice, is it truly evil?

If, as pointed out in Chapter 2, our virtual lives follow a cosmic blueprint that unfolds in logical ways in response to the choices we make in life and if these other people in our intangible worlds are spectral, put there simply to show us the choices we may make in life and the consequences of those choices, it is difficult to find evil in the creator of our world and its blueprint. Just as we do not judge to be evil the producer of a movie or play that depicts evil, we should not judge to be evil the creator of this world for allowing virtual evil. We no longer have a "problem of evil."

---

[16] "The Problem of Evil," Stanford Encyclopedia of Philosophy, 2009.

We Are Not Alone

A final reason that I suggest that the purpose of life is to choose between good and evil is that it provides a possible answer to the question of who these other people in our worlds are and why they are here, sharing our worlds. And when I speak of these other people, I am referring to two groups of people: first, those people who occupy our everyday world and live with us in our universe, and second, our counterparts in other of the many universes of our multiverse. Both of these groups of people, in the dual-reality framework, come into existence with our awareness at birth and go out of existence when we die.

Speaking for the moment about that first group of people, if this everyday world is all in our minds, as I have shown it almost certainly is, why does it contain other people who also are only in our minds? Are these other people, and perhaps all living creatures, our clones in some sense so that we may count on their helping us both to learn about the good or evil in ourselves and to assist us in carrying out our choices of good or evil?

Anyone conversant with current events and world history cannot avoid the observation that one of the most obvious characteristics of the world is its astonishing array of good and evil. The range encompasses very noble acts of courageous self-sacrifice and acts of almost unbelievable cruelty. And these instances of good and evil appear in a seemingly endless variety of types and forms. Further, it frequently is difficult to distinguish between good and evil, especially the good or evil in ourselves, for each seems to have many disguises. The world I have just described can be seen as, and I believe is, a remarkable classroom for learning about good and evil. By observing and interacting with the other people of our worlds, are we to learn about ourselves? If, as suggested in Chapter 1, and as the late John Wheeler suggested, the world is made of *information*, is it not logical to infer that the purpose of our having other people in our virtual worlds is to enable our *learning* about the world?[17]

And that second group of people, our counterparts in other universes, why are they there? Perhaps they are the scorekeepers for our acts of good and evil. Choices we make in life cause successive branchings of life and with each branching, we create a counterpart or counterparts (depending on how many choices were available to us) in other universe(s) who chose to follow paths different from our own. Concerned as we appear to be about choices between good and evil, it is especially those choices that may trigger branchings. If we could compare scorecards, so to speak. with our counterparts in other universes, how acceptable were the choices of good and evil we have made in life as compared to those of our counterparts? Given the start we had in life—shared by all of our counterparts—we might find that we have done quite well given where we and they started in life or we might find that, compared to our counterparts, that we have wasted the many opportunities we had to make the world a better place.

And of what ultimate use are these counterparts? Perhaps they serve as our consciences. Or are they to be used as unimpeachable

---

17 "Does the Universe Exist if We're Not Looking?" interview with John Wheeler by Tim Folger, photography by Dan Winters, *Discover*, June 1, 2002.

evidence in a final judgment day that many religions contend awaits us all?[18]

Chapter Summary

The new ways to view reality and time suggested in this book invite a new look at some deep-rooted questions. What is life and the universe—that stage upon which our lives are displayed? Is there a creator? Do our lives have purpose? If so, what is that purpose?

---

[18] For an interesting speculation about a possible purpose of the many-worlds counterparts—to bestow immortality on human beings—see Jérémie Harris, "Quantum Immortality and the Many Lives of Schrödinger's Cat: Forget the Afterlife—Some Physicists Think You Might Already Be Immortal," *Skeptic Magazine*, vol. 19, no. 3, 2014, 42-45.

CHAPTER 6
EPILOGUE; CONCLUDING COMMENTS

If you cannot love unconditionally, you are not qual-
ified to judge. (Donald W. Jarrell)[1]

---

[1] One of the author's own writings entered February 4, 2011, in an unpublished private collection of favorite poems and quotations.

Suddenly, time ending in beginning
of one eternal perfect day for souls. Bertha C. Bays[2]

Writing this book has been a personal discovery process and an emotional experience for me. At an early stage in the process, I felt a sense of loss when I realized at a deep level that the physical objects I loved—especially the things of our natural environment—were not real in the way I previously understood them to be. Over time, however, I came to see my world in a new way and now see that the things of my world that I loved are still here in much the same way as before. And the world I now see is a much more dependable one and more receptive to positive change than the old world. If you have decided that the logical journey I have taken you on has brought you to a more accurate way to view our world, then you may feel, as I do, that :

The new reality is not just a more accurate view, it is
a *better reality*.

We can be better people, and organized religion
could, but often does not, play a constructive role
in helping us in that task.

The new reality suggests that an afterlife is our ultimate destiny.

The New Reality Is Better

Let me remind you what the old reality for most of us, including myself, was. Our world was one of "real" physical objects,

---

[2] Bertha C. Bays, *My Beautiful Autumn* (Luverne, AL: Louise Fox, 2003), 28.

and among them we felt sheltered from those countless things, including entropy and death, that are arrayed in opposition to living things. The solid feel of the physical objects in our world was reassuring. It suggested permanence and was an inherent part of the human experience—solidly built things typically last longer.

The problem with this view of reality was that, at a deep underlying level, we knew that our everyday world was not eternal. The imminence of our own deaths and the fragility of our planet were accepted facts in our logical frameworks. And many of us also were aware that science had shown that time and our everyday world itself were illusions. In short, our old reality was a ready prey to examination; it could survive only within a tent of denial.

You may be saying at this point, "But how can an intangible everyday world be an improvement on what we had before." Here is how: The new view of a dual-reality world confronts directly the evidence about our everyday world and gives us a way to view our world that removes much of the opacity of the older view of reality. As a result, we can now see at least two positive aspects of our everyday world that were not observable before.

First, and perhaps the most important point: it is now evident that our everyday life *is indeed* real. To repeat much of the argument of Chapter 1 above, while it is correct to say that our everyday world is "virtual," it is not "made up." It *happens* to us. And if we stop to really look at the world that is happening to us, we realize how remarkable it is that the world *is*, that we are here, that we can *experience it*. Moreover, while our world follows an inherently logical path into the future, it frequently causes us to say, "You couldn't make this up;" it is an exciting world!

Second, as I also argued in Chapter 1, this new way to view reality and time may lead to a very exciting time of extensive creativity driven by new approaches to learning and research. Perhaps the most important implication will be for the future of science itself and its role in the advancement of knowledge. Knowing that we live in a world of two realities and that logic works in only one of these realities, we may need a new profession beyond those of physics and philosophy that pursues a new level of understanding beyond rational explanation. Its practitioners would need to be skilled at the practice of meditation and would give emphasis to thought experiments as a way to explore the frontiers of knowledge, while not abandoning research skills and physical experiments.

From within this context, we could examine the role that mystics and meditation might play in further extending our view of reality. Why do mystics typically view what they experience during meditation as more real than their everyday lives? And why are mystics changed by the practice of meditation, achieving a state of psychological well-being substantially higher than the national average?[3] We need more research and study of the mystical experience to see what role it may hold for ordinary citizens.

We Can Be Better People

As pointed out above, the choice between good and evil is something we inherently do. Moreover, we seem to be meant, in

---

[3] Shapiro, Walsh, and Britton, "An Analysis of Recent Meditation Research and Suggestions for Future Directions", 87.

some sense,[4] to engage in spiritual activity, and during this activity we appear to engage with something beyond ourselves and our world.

Whatever our religious beliefs, we can do a better job of choosing to do good and doing away with evil both by taking individual action and by banding together to support activities directed to the end of improving the quality of life for all humankind. Getting agreement on the structure and operational details of an activist group can be difficult but certainly is possible, especially with the wealth of information available on the Internet to assist people interested in forming an activist group.[5] A way to accomplish this end in the past was to start a new religion. There is reason to question whether this action will lead to better people and a better world.

A Role for Organized Religion?

Can organized religion play a constructive role in promoting the common good of humankind? The track record of religion in this regard is not good. It frequently has instigated prejudice, hatred of others, and war and must be seen as a cause of many of the world's problems today.

Many persons apparently need a god image in their minds to help them sort out good and evil. Adoption of an image of god shared by others allows people to form religions and to work together to sort out good and evil and to carry out coordinated

---

[4] Andrew Newberg, Eugene D'Aquili, and Vince Rause, *Why God Won't Go Away: Brain Science and the Biology of Belief* (New York, NY: Ballantine Books, 2001), 172.

[5] Enter the exact phrase "forming an activist group" in an Internet search engine of your choice.

action thought to be pleasing to their god. The god-images that the members of these groups serve appear clearly to be in large part a response to the need of individual members of the group to choose between good and evil, for all the world's major religions have doing good and avoiding evil as their primary focus. That this coordinated action meets a need for religious adherents is evidenced by the numbers of people who belong to one of the major religions and by the extent to which membership in a religious group commands the loyalty of its members—a loyalty that frequently rivals or exceeds their loyalty to nations and tribal groupings.[6]

For many people, participation with others in religious ceremonies also appears to enhance a feeling of communion with a creator, to create a "thin place" that enables them to more readily have mystical or transcendent experiences. A thin place in this context is any place where the distance between worshipers and a transcendent being seems especially narrow and worshipers more readily undergo spiritual or epiphanic experiences. A thin-place sensation may occur in a particular place such as a building for religious worship or a natural setting such as a wooded forest, or it may occur when hearing a particular musical piece or seeing a work of art, or it may happen in the company of a particular person or group of persons. To many persons, this appears to be a primary reason for

---

[6] See Daniel Druckman, "Nationalism, Patriotism, and Group Loyalty: A Social Psychological Perspective", *Mershon International Studies Review* (1994) 38, 43-68.

participating in a religious activity, although a thin place-sensation may occur in numerous other contexts as well.[7]

The god-images people have in their minds and the actions these gods require often differ markedly. Consider, for example, the differences between what god is felt to require of believing Christians, Jews, and Muslims or even the wide differences between the gods of persons within these religions. These differences suggest that we as humans can decide the image of god that we allow to occupy our minds and, in turn, to dictate our choices of good and evil. If the purpose of *individual* life is to choose between good and evil, then those persons who use a god-image to define good and evil must consider carefully the image of god they allow to enter their minds. A self-serving god, for example, who designates followers as preferred people and approves without exception activities that the followers themselves choose to undertake is not serving the purpose of helping individual members of the group to choose between good and evil and is unlikely to better life for all humankind. Organized religion can play a constructive role in our lives only if its adherents become more critical of the religious groups they choose to join and of the god-image they choose to provide focus for their lives. The potential benefit of organized religion for humankind will not be realized unless we demand more from our religions than we have in the past.

---

[7] For a discussion of the role of thin places in Christianity, see Marcus J. Borg, *The Heart of Christianity: Rediscovering a Life of Faith* (New York, NY: HarperCollins, 2003), 149-163.

A Speculation: Is There an Afterlife?

The dual-realities view of reality offers several reasons for thinking there may be an "add on" of some kind beyond our lives here on Earth. First, our individual everyday lives have been shown to very likely have a purpose. Second, each individual human life appears to have been brought into being by the investment of an incredible amount of resources. Finally, each of our daily lives is maintained by a continual creation process and is guided by a cosmic blueprint. These bits of evidence suggest that the lives of *individuals* are not chance events but are created, valued, and maintained for a reason.

In purely human terms, it seems unlikely that all of this "investment" would be wasted by simply ending the story of our lives at our deaths. And since the focus in this life (in the dual-realities view) is on the individual, it seems likely that *individual identity* will be preserved in an afterlife. My conclusion: we have an amazing and marvelous afterlife ahead of us.

BIBLIOGRAPHY

Alley, C. O., Jakubowicz, O., Steggerda, C. A. and Wickes, W. C. "A Delayed Random Choice Quantum Mechanics Experiment with Light Quanta." In Kamefuchi, S. et al. (eds.). *Proceedings of the International Symposium on the Foundations of Quantum Mechanics*, Tokyo, 1983 (Tokyo, Japan, Physical Society of Japan, 1984).

Armstrong, Karen. *A History of God: The 4000-Year Quest of Judaism, Christianity and Islam* (New York, NY: Ballantine Books, 1993).

Arnsten, Amy, Mazure, Carolyn M., and Sinha, Rajita. "This Is Your Brain on Meltdown," *Scientific American*, volume 306, no. 4, April 2012, 48-53.

Aspect, Alain and Grangier, Philippe. "Experiments on Einstein-Podolsky-Rosen-type Correlations with Pairs of Visible Protons," 1-15, in Penrose, R. and Isham, C. J. *Quantum Concepts in Space and Time* (Oxford, Eng.: Clarendon Press, 1986)

Baez, John, Unruh, William G., and Tifft, William G. "Is Time Quantized? In Other Words, Is there a Fundamental Unit of Time That Could Not Be Divided into a Briefer Unit?," *Scientific American*, Thursday, October 21, 1999, 12.

Baron-Cohen, Simon. "Autism - 'Autos': Literally, a Total Focus on the Self?," 166-180, in Feinberg, Todd E. and Keenan, Julian Paul (eds.). *The Lost Self: Pathologies of the Brain and Identity* (Cambridge, UK: Oxford University Press, 2005).

Barrow, John D. *The Constants of Nature: From Alpha to Omega-the Numbers That Encode the Deepest Secrets of the Universe* (New York, NY: Pantheon Books, 2002).

Barrow, John. *Theories of Everything: The Quest for Ultimate Explanation* (Oxford, Eng.: Oxford University Press, 1991).

Bechara, Antoine, Damasio, Hanna, Tranel, Daniel, Damasio, and Antonio R. "Deciding Advantageously before Knowing the Advantageous Strategy." *Science*, 275, 1997, 1293-95.

Bekenstein, Jacob D. "Information in the Holographic Universe: Overview -The World as a Hologram," *Scientific American*, July 14, 2003.

Brown, George I. and Gaynor, Donald. "Athletic Action as Creativity," *Journal of Creative Behavior*, vol. 1, no. 2 (1967).

Byrne, Peter. "The Many Worlds of Hugh Everett," *Scientific American*, December 2007, 98-105.

Byrne, Peter. Interview with Leonard Susskind. "Leonard Susskind, Bad Boy of Physics," *Scientific American*, July 2011, 80-83.

Cardeña, Etzel. "Anomalous experiences and hypnosis," *Proceedings of the 49th Annual Conventions of the Parapsychological Association*, 2006 and 2007, 32-42.

Cohen, Martin. *Wittgenstein's Beetle and Other Classic Thought Experiments* (Malden, MA: Blackwell Publishing, 2005).

Cooper, Linn F. and Erickson, Milton H. *Time Distortion in Hypnosis; An Experimental and Clinical Investigation* (Boca Raton, FL: OTC Publishing Corp, 2004). (First published in 1954)

Darling, David. *The Universal Book of Mathematics: From Abracadabra to Zeno's Paradoxes* (Google eBook) (New York, NY: John Wiley & Sons, Aug 11, 2004).

Davies, P. C. W. "Where Do the Laws of Physics Come From?" www.scribd.com/doc/.../ Where-Do-the-Laws-of-Physics-Come-From, Oct 8, 2008.

Davies, P. C. W. and Brown, J. R. *The Ghost in the Atom: A Discussion of the Mysteries of Quantum Physics* (Cambridge, Eng.: Cambridge University Press, 1986).

Davies, Paul. "The Mysterious Flow of Time," *Scientific American*, Sep. 2002, 32-37.

Davies, Paul. *The Cosmic Blueprint: New Discoveries in Nature's Creative Ability To Order the Universe* (New York, NY: Simon and Schuster, 1988).

Davies, Paul. *The Mind of God* (New York, NY: Simon and Schuster, 1992).

Descartes, René. *Discourse on Method and Meditations*. Translated with an Introduction by Laurence J. Lafleur (New York, NY: Macmillan Publishing Co., 1960) (Originally published in 1637).

Deutsch, David. "Three Connections between Everett's Interpretation and Experiment," 215-225, in Penrose, R. and Isham, C. J. *Quantum Concepts in Space and Time* (Oxford, Eng.: Clarendon Press, 1986).

Deutsch, David. *The Fabric of Reality: The Science of Parallel Universes and Its Implications* (New York, NY: The Penguin Press, 1997).

Eagleman, David. "The Secret Life of the Mind," *Discover*, September 2011, 50-53.

Einstein, Albert and Infeld, Leopold. *The Evolution of Physics; The Growth of Ideas from Early Concepts to Relativity and Quanta* (New York, NY: Simon and Schuster, 1961). (Original copyright, 1938).

Erickson, M. H. "Deep hypnosis and its induction," 70-114 in LeCron, L. M. (ed.). *Experimental Hypnosis* (New York, NY: Macmillan, 1952).

Everett, Hugh, III. *The Many-Worlds Interpretation of Quantum Mechanics: The Theory Of The Universal Wavefunction* (Doctoral Dissertation, Princeton University, 1957). May be downloaded as a pdf file at www.pbs.org.

Feinberg, Gerald. *What Is the World Made of: Atoms, Leptons, Quarks, and Other Tantalizing Particles* (Garden City, NY: Anchor Press, 1977).

Feynman, Richard P. *QED: The Strange Theory of Light and Matter* (Princeton, NJ: Princeton University Press, 1985).

Fink, Andreas, et al. "The Creative Brain: Investigation of Brain Activity during Creative Problem Solving by Means of EEG and FMRI," *Human Brain Mapping*, March 2009, Volume 30, Issue 3.

Folger, Tim. Interview with John Wheeler, "Does the Universe Exist if We're Not Looking?", *Discover*, June 1, 2002.

Gaidos, Susan. "The Mesmerized Mind," *Science News*, October 10, 2009, Vol. 176, Issue 8, 26-29.

Ghiselin, Brewster (ed.). *The Creative Process: A Symposium* (New York, NY: New American Library, 1954).

Giese, Albrecht. *Origin of Gravity*, available at http://ag-physics. org/gravity.

Greeley, A. "Mysticism Goes Mainstream," *American Health*, Vol. 6 (1), 1987, 47-49.

Green, Elmer, and Green, Alyce. *Beyond Biofeedback* (New York, NY: Delacourte Press, 1977)

Greene, Brian. *The Elegant Universe: Superstrings, Hidden Dimensions, and the Quest for the Ultimate Theory* (New York, NY: W. W. Norton and Company, 1999).

Greene, Brian. *The Fabric of the Cosmos: Space, Time, and the Texture of Reality* (New York, NY: Alfred A. Knopf, 2004).

Greene, Brian. *The Hidden Reality: Parallel Universes and the Deep Laws of the Cosmos* (New York, NY: Alfred A. Knopf, 2011).

Greenemeier, Larry. "Holographic Film for 3-D, sans Those Silly Specs," *Scientific American*, March 3, 2008,

Gribbin, John. *Schrödinger's Kittens and the Search for Reality* (Boston, MA: Little, Brown and Company, 1995).

Harris, Jérémie. "Quantum Immortality and the Many Lives of Schrödinger's Cat: Forget the Afterlife—Some Physicists Think You Might Already Be Immortal," *Skeptic Magazine*, vol. 19, no. 3, 2014, 42-45.

Hawking, Stephen W. *A Brief History of Time: From the Big Bang to Black Holes* (New York, NY: Bantam Press, 1988).

Heise, David R. *Surveying Cultures: Discovering Shared Conceptions and Sentiments* (Hoboken NJ: Wiley, 2010).

Himmelfarb, S. "The measurement of attitudes," 23-87, in A.H. Eagly & S. Chaiken (eds.) *Psychology of Attitudes* (New York, NY: Harcourt Brace Jovanovich, 1993).

Hinshaw, Karin E. and Hansen, Jeffrey Jay. "The Effects of Mental Practice on Motor Skill Performance: Critical Evaluation and Meta-Analysis," *Imagination, Cognition and Personality*, Volume 11, Number 1, 1991-92, 3-35.

Hunt, Amelia. "An Interesting 100 Milliseconds in Which To Examine Vision and Attention," a seminar presentation at the Vision Sciences Lab, Harvard University, Wednesday, March 12, 2007, cited by Zimmer, Carl. "How Your Brain Can Control Time," *Discover*, July 12, 2008.

Janssen, Jeffrey J. and Sheikh, Anees A. "Enhancing Athletic Performance through Imagery: An Overview," 1-22, in Sheikh, Anees A. and Korn, Errol R. (eds.). *Imagery in Sports and Physical Performance: Imagery and Human Development Series* (Amityville, NY: Baywood Publishing Company, 1994).

Johnson, George. *A Shortcut through Time: The Path to the Quantum Computer* (New York, NY: Alfred A. Knopf, 2003).

Johnson, George. *Fire in the Mind: Science, Faith, and the Search for Order* (New York, NY: Alfred A. Knopf, 1995).

King, Billie Jean with Chapin, Kim. *Billie Jean* (New York, NY: Harper and Row, 1974).

LeShan, Laurence and Margenau, Henry. *Einstein's Space and Van Gogh's Sky* (New York, NY: Macmillan, 1982).

Libet, B., Gleason, C. A., Wright, E. W., and Pearl, D. K. "Time of Conscious Intention To Act in Relation to Onset of Cerebral Activity (readiness-potential): The Unconscious Initiation of a Freely Voluntary Action." *Brain*, 106, 1983, 623-642.

Maldacena, Juan. "The Illusion of Gravity," *Scientific American*, November 2005, 56-57.

Manso, Peter. *Vrooom!!: Conversations with the Grand Prix Champions* (New York, NY: Funk and Wagnalls, 1969), 180-181.

Margenau, Henry. *The Nature of Physical Reality: A Philosophy of Modern Physics* (New York, NY: McGraw-Hill, 1950).

Maslow, A. H. "The Good Life of the Self-Actualizing Person," 46-53, in Covin, Theron M. (ed.). *Readings in Human Development: A Humanistic Approach* (Beverly Hills, CA: MSS Publishing Corp., 1974).

Morris, Richard. "Inventing the Universe." In John Brockman (ed.), *Creativity* (New York, NY: Touchstone, 1993), 130-148

Murphy, Michael and White, Rhea A. *The Psychic Side of Sports* (Reading, MA: Addison Wesley, 1978).

Nadis, Steve, "Time Asymmetry Finally Found," *Discover*, June 2013, 14.

Naish, P. L. "Hypnotic Time Perception: Busy Beaver or Tardy Timekeeper?," *Contemporary Hypnosis*, 2001, 18, 87-99.

Newberg, Andrew, D'Aquili, Eugene, and Rause, Vince. *Why God Won't Go Away: Brain Science and the Biology of Belief* (New York, NY: Ballantine Books, 2001).

Nicklaus, Jack with Bowden, Ken. *Golf My Way: The Instructional Classic Revised and Updated* (New York, NY: Simon and Schuster, 2005).

Osgood, C. E., Suci, G., & Tannenbaum, P. *The Measurement of Meaning* (Urbana, IL: University of Illinois Press, 1957).

Penrose, Roger. *The Emperor's New Mind: Concerning Computers, Minds and the Laws of Physics* (Oxford University Press, Oxford, 1989).

Phenix, Philip H., "Good and Evil," Chapter 9 in *Intelligible Religion* (New York: Harper & Brothers, 1954), 72-62.

Pring, L. and Hermelin, B., "Numbers and Letters: Exploring an Autistic Savant's Unpracticed Ability," *Neurocase*, 2002, 8(4), 330-337.

Quinn, Helen R., "Time Reversal Violation," Talk presented at the Discrete '08 Conference, and published in the *Journal of Physics Conference Series*, p. 3.

Ravizza, Kenneth. "Peak Experiences in Sport," *Journal of Humanistic Psychology*, 17: 1977, 35-40, reprinted in Smith, Daniel and Bar-Eli, Michael. (eds.). *Essential Readings in Sport and Exercise Psychology* (Champaign, IL: Human Kinetics, 2007), 122-125.

Rees, Martin. *Just Six Numbers: The Deep Forces that Shape the Universe* (New York, NY: Basic Books, 2000).

Richardson, Alan. "Mental Practice: A Review and Discussion, Part I," *Research Quarterly*, vol. 38, 1967, 95-107.

Richardson, Alan. "Mental Practice: A Review and Discussion, Part II," *Research Quarterly*, vol. 38, 1967, 263-273.

Russell, Bertrand. "Knowledge by Acquaintance and Knowledge by Description," in Egner, Robert E. and Denonn, Lester E. (eds.). *The Basic Writings of Bertrand Russell, 1903-1959* (New York, NY: Routledge, 1961), 217-224.

Sachdev, Subir. "Quantum Physics: Strange and Stringy." *Scientific American*, January 2013, 44-51.

Seife, Charles. *Decoding the Universe: How the New Science of Information is Explaining Everything in the Cosmos, from Our Brains to Black Holes* (New York, NY: Penguin Group (USA) Inc., 2006).

Shapiro, Shauna L., Walsh, Roger, and Britton, Willoughby B., "An Analysis of Recent Meditation Research and Suggestions for Future Directions," *Journal for Meditation and Meditation Research*, Vol. 3, 2003, 69-90.

Sheikh, Anees A. and Korn, Errol R. (eds.). *Imagery in Sports and Physical Performance: Imagery and Human Development Series* (Amityville, NY: Baywood Publishing Company, 1994).

Soon, Chun Siong, Brass, Marcel, Heinze, Hans-Jochen, and Haynes, John-Dylan. "Unconscious Determinants of Free Decisions in the Human Brain," *Nature Neuroscience* (Nature Publishing Group) 11 (5), April 13, 2008, 43-45.

St Jean, R., McInnis, K., Campbell-Mayne, L., and Swainson, P. J. "Hypnotic Underestimation of Time: the Busy Beaver Hypothesis," *Journal of Abnormal Psychology*, August 1994,103 (3), 565-9.

Start, Kenneth B. and Richardson, Alan. "Imagery and Mental Practice," *British Journal of Educational Psychology*, vol. 34, no. 3, 1964, 280-284.

Stetson, C., Fiesta, M. P., and Eagleman, D. M. "Does Time Really Slow Down during a Frightening Event? PLoS ONE, 2(12), 2007.

Tegmark, Max. "The Interpretation of Quantum Mechanics: Many Worlds or Many Words?," September 15, 1997. Available online at sns.ias.edu/~max/everett.html.

Tse, Peter Ulric, Intriligator, James, Rivest, Josée, and Cavanagh, Patrick. "Attention and the Subjective Expansion of Time," *Perception & Psychophysics*, 2004, 66 (7), 1171-1189.

Underhill, Evelyn. *Mysticism: The Nature and Development of Spiritual Consciousness* (Oxford, Eng.: Oneworld Pubs., 1993). (First published in 1911).

Van Flandern, T. "What the Global Positioning System Tells us about Relativity," 81-90, in Selleri, F. (ed.). *Open Questions in Relativistic Physics* (Montreal, Can: Apeiron, 1998).

Waldron, Joan L. "The Life Impact of Transcendent Experiences with a Pronounced Quality of Noesis," *The Journal of Transpersonal Psychology*, 1998, Vol. 30, No. 2, 103-134.

Wheeler, John Archibald. "The 'Past' and the 'Delayed-Choice' Double-Slit Experiment." In Marlow, A. R. (ed.) *Mathematical Foundations of Quantum Theory* (New York, NY: Academic Press, 1978).

Wigner, Eugene P., "The Unreasonable Effectiveness of Mathematics in the Natural Sciences." In Ferris, Timothy (ed.). *The World Treasury of Physics, Astronomy, and Mathematics* (Boston, MA: Little, Brown, 1991).

Woolfolk, Robert L., Parrish, Mark W., and Murphy, Shane M. "The Effects of Positive and Negative Imagery on Motor Skill Performance," *Cognitive Therapy and Research*, Vol. 9, Issue 3, June 1985, 335-341.

Zimbardo, P. G., Marshall, G., and Maslach, C. "Liberating Behavior from Time-bound Control: Expanding the Present Through Hypnosis," *Journal of Applied Social Psychology*, 1971, 1, 4, 305-323.

Zimmer, Carl. "The Brain: Stop Paying Attention: Zoning Out Is a Crucial Mental State: Researchers Say a Wandering Mind May Be Important to Setting Goals, Making Discoveries, and Living a Balanced Life," *Discover*, June 15, 2009.

Zimmer, Carl and West, Jeff. "Whose Life Would You Save?: Scientists Say Morality May Be Hardwired into our Brains by Evolution," (Interview with philosopher and neuroethicist Joshua Greene) *Discover*, April 21, 2004, 2-4.

Zohar, Danah. "Creativity and the Quantum Self," 202-218, in Brockman, John (ed.). *Creativity* (New York, NY: Touchstone, 1993).

Zukav, Gary. *The Dancing Wu Li Masters: An Overview of the New Physics*, hardcopy edition (New York, NY: William Morrow and Company, 1979)

INDEX

absolute rest frame,
 75, 131-132
absolute time, 75, 132
afterlife
  and the dual-reality view of
   reality, 152
  is there an, 158
alternative history clusters,
 40, 65-66
  and autism, 65-66
  and schizophrenia, 65
  as normality, 65-66
Anonymous, 136
artists

and meta-logical thinking, 41
and the logical imperative, 40-41
as creative thinkers, 35
discoveries from far and near,
 37-40
mathematicians as artists, 41
role in expressing truth, 40-41
Aspect experiment, 117-118
  and holographic movie model
   of time, 116-118
  and "spooky action at a
   distance", 117
  simultaneity of the two
   measurements, 117

autism

and alternative history clusters,

65-66

autistic savant, 122

Bays, Bertha C. , 152

Bechara, A., 53-54

author suggestion to modify

research design, 54-55

Bekenstein, Jacob, 72

Bell's Theorem

and principle of local causation, 77

holographic movie model and,

116-118

Börne, Ludwig , 135

brain, human

complexity, 6, 8

how it functions, 5-6

studies of, 6-7

subconscious mind as source of

ideas?, 62-63

Buddha, 1

choosing between good and evil

and Hebrew bible, 142-144

and world religions, 142

role for organized religion?,

155-157

Cohen, Martin

and thought experiments, 36-37

comparative time, see time,

comparative

condensed-matter research

and particles in superposition,

122-123

continual creation

and cosmic blueprint, 77-79

and holographic movie model,

75-79

and movement through space,

76-77

defined, 75-76

demonstrated by delayed choice

experiment, 75

continual creation, consequences of,

75-79

a cosmic blueprint required, 77-79

we move through space, not time,

76-77

continual creation, proving, 100-103

each frame a new creation,

102-103-

finding an indivisible unit of

space, 100-102

Cooper, Linn F., 43-44, 45, 46,

90, 92, 94, 98, 99, 106,

107-109

cosmic blueprint

and causal relationship between
time frames, 77
and free will, 77-78
and massive objects, 130
and purposeful life, 141
and response to life choices, 78-79
subject to the logical imperative, 78
time-stamped fermions as
code in, 129
creative performers, 24
Brodie, John, 88
compared to creative thinkers,
46-47
Evans, Lee, 49-50
King, Billie Jean, 48-49
Nicklaus, Jack, 49
Wigman, Mary, 47-48
work habits of, 46-47
creative performing
and kinesthetic learning, 47
discoveries of ideas and images,
34-35
use of imagery, 47-50
creative thinkers
compared to creative performers,
46-47
work habits of, 33-34
creative thinking, discovery of ideas

and the thought experiment, 36-37
by artists and scientists compared,
40-41
Einstein, Albert, and, 35
from far and near, 37-38
in subconscious memory?,
62-64
in the minds of many-worlds
counterparts , 63-64, 121
Penrose, Roger, and, 35
Plato and, 34-35
creator
and the problem of evil, 147
communion with, 156-157
is there one?, 139-140
logically unable to discuss, 139
need for image of, 155-156
D'Aquili, Eugene, 7, 8, 31
Darwin, Charles
Evolution-Theory Thought
Experiment, 37
ideas ahead of his time, 38
Davies, Paul, 4-5, 10, 59, 72, 77
delayed-choice experiment
a "big bang" event in each time
frame , 111
and a virtual everyday world. 29
and continual creation, 75

and information in other worlds,
29, 52

and the pointillist principle, 29

described, 37-39, 52

Descartes, René, 20

and the information-world

postulate, 20

we exist in thought only, 20

Deuteronomy 30:15-19, 142-143

Deutsch, David, 30, 52-53, 66-67, 120

discoveries of innovative ideas

and images

and certainty by discoverer about,
56-57

and the thought experiment,
36-37

by artists and scientists compared,
40-41

by performers, 50

from far and near, 37-40

in the subconscious mind?, p.
62-63

through transcendental experience,
34-35

double-slit experiment (see two-slit
experiment)

dual-reality interpretation

a better reality, 152-154

allows more complete view of

our world, iv

and the enigmatic ideas of

science, 51

and the likelihood of an

afterlife, 158

and the need for a new

profession, 59-60

contrasted with the older

view of reality, 152-154

defined, iii, 9, 50, 51

effect of disclosure on everyday

life, 58

implications of, 58-61

importance of individual in, iv,
136, 138, 158

possible impact on science. 58-61

dual-reality interpretation,

proving, 40-45

and macro-world research, 53-55

and quantum-level research, 52-53

and the delayed-choice

experiment, 52

and the "extra body" problem,
55-56

a proposed experiment, 54-55

by showing information available

in other worlds, 51-55

can it be proven?, 50-56

conclusions subject to testing
by experiment, iv

definitive proof not possible, iii,
50-51

Deutsch thought experiments,
52-53

experienced truth an alternative to
experimental proof, 56-58

indirect evidence, 51

proof and the "extra-body"
problem, 55-56

proof through application, iii-iv

Eddington, Sir Arthur Stanley
espousal of mysticism, 59

Einstein, Albert 1, 9, 35, 37, 38, 56,
59, 63, 75, 117

discovery of ideas, 56

espousal of mysticism, 59

Light Wave Thought Experiment, 37

physical concepts as creations of
the mind, 1, 35

relativity theories ahead of their
time, 63

"spooky action at a distance", 117

theory of everything beyond reach
of science, 9-10

elementary particles

and the pointillist principle, 17-19

as bits of information only, 4-5

as building blocks of our
virtual world, 18-19

change behavior when
watched, 4

informing us about our
everyday world, 18

secondary, derivative nature, 5

Emerson, Ralph Waldo, 1

entropy
and increasing complexity of
everyday lives, 66-67

and increasing differentiation of
multiverse, 66-67

an implication of information-
world postulate, 21

defined, 21, 66

epiphanic experiences
religion and, 156

thin places and, 156-157

Erickson, Milton, 98

Evans, Lee, 49-50

Everett branches of multiverse,
24, 118

and our counterparts in other
universes, 149

reality of, 24

Everett, Hugh, III, 23, 24, 63, 120, 123
Everett postulate
  and separation of many-worlds
    histories, 30
everyday world
  and classical reality, 2
  and the information-world
    postulate, 20-21
  and the logical imperative, 21-22
  and the many-worlds
    interpretation, 22-25
  and the pointillist principle, 17-19
  an intangible illusion, ii
  argument for reality of, 8-9, 153
  as one of the dual-realities, ii-iii,
    9, 50
  assembled in the brain, 7
  change occurs between movie
    frames, 81-82, 117, 118
  implications of its being
    intangible, 17-22
  increasing complexity, 66-67
  inherently logical, 21-22
  knowing, 17-50
  looking beyond, 10-17
  made of information, not matter, 20
  organizing principle used to
    present it to us, 22

  projected from beyond time and
    space, 11, 72, 126-127, 139
  what is our everyday world?, 22-25
Exodus 4-15, 144
fermions
  and Pauli exclusion principle, 129
  projected one at a time, 129
  unique time stamp, 129
  use as code in cosmic blueprint, 129
Feynman mirror experiment
  and alternative history clusters,
    39-40
  discoveries in nearby histories, 40
  time projected in many
    dimensions 110-111
Feynman, Richard P., 39
free fall
  defined, 132
  numerical equivalence of gravity
    and inertia, 126, 132-133
free will
  and Heisenberg uncertainty
    principle, 78
  cosmic blueprint responds to our
    choices, 78-79
  is relative in our everyday world, 78
fundamental frame rate
  and absolute time and rest frame, 75

defined, 74-75

Galileo, Galilei

    Gravitational Balls Thought

        Experiment, 37

    scientific works ahead of his

        time, 37-38

Genesis 18:16-33, 143

Giese, Albrecht

    gravitational time dilation at the

        quantum level, 131-132

gravity, see gravitational force

gravitational attentional-focus time-

        distortion effect

    caused by constricted focus of

        attention, 95-96

    described, 95-96

    during hypnotic time distortion,

        98-99

    during impending-peril and

        peak-experience, 96-97

gravitational force

    and mass, 130-131

    and time-stamped fermions,

        128-129

    defined, 126

    in holographic movie model,

        128-131

    linkage with inertia, 128-133

linkage with time, 128-131

    numerical equivalence with inertial

        force, 132-133

    Pauli exclusion principle, 129

Greene, Brian, 72

Guth, Alan

    ideas ahead of his time 38

habeas corpus

    in proof of time distortion,

        106-109

Hawking, Stephen, 38

    ideas ahead of his time 38

    time began with the "big bang", 111

Heisenberg, Werner Karl

    espousal of mysticism, 59

    uncertainty principle, 2-3, 78

Heraclitus, 70, 115

Hermelin, B., 122

holographic film, 73-74

holographic movie model of time

    and a fundamental frame rate, 75, 83

    and Bell's Theorem, 77, 116-118

    and moving through space at speed

        of light, 74-75, 76-77, 127

    and research about particles in

        superposition, 121-124

    and the many-worlds

        interpretation, 118-121

change occurs between time
frames only, 81, 117, 118
described, 72-75
displays our world in many
dimensions, 110-111
each frame a new creation,
102-103
each frame contains a "big bang"
event, 111-113
movement can only slow time,
81, 127
proving, 100-109
proving a key concept: each frame
a new creation, 102-103
proving a key concept: personal
time , 103-109
proving through successful
application, 100
ultimate proof or disproof not
possible, 100
holographic principle
and bringing past into present,
112-113
defined, 19, 112-113
universe subject to, 18-19, 112
holographic projection
Brian Greene concerning, 72
Jacob Bekenstein concerning, 72

holographic theory
substantiates the pointillist
principle, 19
how to define for universe, 113
holographic universe, 18-19, 112-113
human bodies, as computer
programs, 64
implications for healthcare, 64
Hunt, Amelia, 53
hypnosis, self-induced, 49
hypnosis, time distortion
during, 89-94
dress-design experiment, 91-92
experimental procedure, 90-92
gravitational effect on time
passage, 98-99
other studies, 92-94
procedure for performing
hypnosis, 90
subjects feel time passes normally,
89-90, 99
"watched pot" effect not a factor, 99
hypnosis, to enhance creativity, 42-46
dress-design experiment, 43-45
experimental procedure, 43-45
origin of new ideas, 45
performance compared to waking
experience, 45

procedure for performing
  hypnosis, 43
hypnotic trance
  and the "extra-body" problem,
    55-56
  comparison with self-induced
    trance, 42
  neuroimaging study of brain
    during, 42
  reality of experience, 46
impending peril, time distortion
  during, 86
  accompanied by intense focus
    on events, 86
  gravitational effect a likely
    factor, 96-97
  sensation of slowing time, 86
  survival value of, 86
  tangible proof of time distortion
    possible?, 104-109
  time passes more rapidly for
    observers, 97
  "watched pot" effect a likely
    factor, 96-97
inertia, see inertial force
inertial force
  and continual creation, 127-128
  defined, 125-126

in holographic movie model,
    126-128
  linkage with gravity, 128-132
  linkage with time, 128-131
  movement results in, 128
  numerical equivalence with
    gravitational force, 132-133
Infeld, Leopold, 9, 35
information world postulate, 20-21
  and entropy, 21, 66-68
  and John Wheeler, 20
  and René Descartes, 20
  an implication of the virtual-world
    assumption, 20-21
  defined, 20
Jarrell, Donald W., 125, 151
Jeans, Sir James Hopwood
  espousal of mysticism, 59
Job 1:1-22, 144
kinesthetic learning
  defined, 47
  importance for performance, 50
King, Billie Jean, 48-49
knowing everything, limitations on
  and information beyond time and
    space, 9-11
  and information in other multiverse
    worlds, 10, 25-29, 121-123

and our limited reasoning
capacities, 15-16
and wave function collapse, 26
knowledge inherently incomplete,
Einstein and Infeld, 9-10
scientific knowledge incomplete,
Davies, 10, 59
knowing our everyday world
and information in other multiverse
worlds, 10, 25-29, 121-123
can we retrieve information from
other worlds?, 30-32
implications of its virtual nature
for knowing it, 17
other worlds contain information
useful to us, 25-29
what is our everyday world?, 22-25
knowing the world beyond time and
space, 10-17
learning from the mystics, 11-17
LeShan, Laurence, 78-79
Lewin, Kurt, i
Libet, Benjamin, 54
life
a purpose for?, 140-142
coexistent with time, 137
result of a miraculous event, 140
what is it?, 137

light
straight-line travel, 39
limited capacity of human mind to
reason
and need to classify, 15-16
logical imperative
and completeness of autobiographical
memory, 108
and logical unfolding of cosmic
blueprint, 112
and remarkable success of
mathematics, 21-22
and the work of artists, 40-41
an implication of a virtual world,
21-22
defined, 22
does not allow logically impossible
histories, 24
in the worlds of hypnotist and
subject, 55-56
loss of information
and an all-encompassing truth, 16
and classification of information,
15-16
and the universal wave
function, 26
and wave function collapse, 26
Mach, Ernst

Motionless Chain Thought
Experiment, 37
many-worlds interpretation of
quantum mechanics
and alternative history clusters,
40, 66
and a theory of everything, 27
and increased complexity of time
and space, 24-25
and information in other
multiverse worlds, 25-29
and wave function collapse, 25-27
contrasted with one-world
view, 24
description, 23-25
reality of many-worlds branches, 24
searching alternative histories in,
25, 30-32
the delayed-choice experiment and
MWI, 27-29
time in the multiverse does not
pass but is, 34
Margenau, Henry, 78-79, 101
Marshall, Gary, 93-94
Maslach, Christina, 93-94
Maxwell, James Clerk
deathbed revelation, 62
Demon Thought Experiment, 37

meditation
antiquity of practice, 11
as practiced by Buddhist monks,
11-13
neurological study of, 12-13
what happens during?, 13-14
mental illness
as difficulty defining reality, 64
autism and alternative histories,
65-66
schizophrenia and alternative
histories, 65
Micah 6:8, 136
Morris, Richard, 38
multiverse
defined, 24
multiverse worlds
all equally real, 24
can we retrieve information
from?, 30-32
contain information useful to
us?, 25-29
our other-world counterparts as
evidence for a judgment?,
149-150
our other-world counterparts
serve as our consciences?,
149

our other-world counterparts—
why are they there?, 149
retrieving information by
transcendent experiences,
33-50, 51
muons
and comparative lifetimes, 84-85
time dilation for non-living
moving things?, 84-85
mystical experience
and the unitary continuum, 31
compared to transcendental
experience, 31-32
objective of, 32
mystics
and connotative learning, 16-17
and peak meditative state, 14, 15,
56
and syntactic learning, 16-17
compared to explorers on Earth,
57
difficulty telling others what they
learn, 14
higher levels of psychological
health, 14-15, 57-58, 154
what they learn while meditating,
14-17
Nadis, Steve, 103

near accident, see impending peril
neuroimaging
and hypnotic effects on brain
function, 42
and our understanding of how the
brain functions, 5-6
challenge to classical view of
reality, 5-8
described, 5-7
studies of mystical experience, 7,
12-13
Newberg, Andrew, et al., 7, 8, 12,13.
30, 31, 32
study challenging classical view of
reality, 7
study of meditation by mystics,
12-13
unitary continuum, 30-31
Newton, Isaac
Bucket Thought Experiment. 37
works ahead of his time, 38
Nicklaus, Jack, 49
observation
and behavior of elementary
particles, 4, 27
and "collapse of the wave
function", 14, 25-26,
137-138

and ongoing creation of past,
present, and future, 75

and particles as abstract
constructions, 5

defined as "to attempt to
observe", 13-14

observe, see observation

observer, role of

creating reality, 20

in many-worlds interpretation, 24

in time distortion, 86, 87-88, 94-99

organized religion

constructive role in promoting
common good?, 155-157-

and epiphanic experiences,
156-157

and our purpose in life, 142-144

and the choice of a god-image,
155-156

and the "problem of evil", 147

differences among god-images
chosen, 157-158

fulfills need for coordinated action
by adherents, 156

meditation as contemplative core
of, 11

often a disappointment in the past,
155=157

role of a god-image, 155-156, 157

other multiverse worlds

can we retrieve information from?,
30-32

contain information useful to us,
25-29

evidence that we get information
from, 52-55

limits on ability to search, 30, 137

our counterparts as evidence in a
judgment?, 149-150

our counterparts as our
consciences?, 149

other people in our universe

here to help us learn about good
and evil?, 148-149

why do they accompany us?, 148

particles

and the pointillist principle,
18, 19

and the watched pot experiment,
94-95

as abstract constructions, 5

as building blocks of our virtual
world, 17-18

as information only, 4-5

not elementary, 4-5

particles in superposition, 19, 111

all appear at same time in our
world, 117-118, 118-120

and Bell's Theorem, 117-118

and condensed-matter physics, 51,
122-123

and creative thinking, 121-122

and future research, 116, 121-123

and quantum computing, 51,
119-121

and the many-worlds
interpretation, 118-121

and time reversal violation,
103-104

as basic units of action, 51,
122-123

defined, 26, 115-116

where are the mirror-image
particles?, 118-119

Pauli exclusion principle
and time-stamped fermions, 129

defined, 129

Pauli, Wolfgang
espousal of mysticism, 59

peak experience, time distortion
during, 87-89

accompanied by intense focus on
events, 87, 96-97

gravitational effect a likely
factor, 97

importance of innate ability and
practice, 87, 89

sensation of slowing time, 87-88

time passes more rapidly for
performers, 96-97

"watched pot" effect a likely
factor, 96-97

peak performance (see peak
experience)

Penrose, Roger
and discovery of mathematical
truths, 35

performing creatively while in self-
induced trance, 46-50

and kinesthetic learning, 47, 87

see also "thinking creatively"

personal time, see time, personal

Planck length
and the thickness of a holographic
movie frame, 73

defined, 73, 101

Planck, Max, 73

Planck time
defined, 73, 101

Plato

and discovery of new ideas and art forms, 34-35

Poincaré, Henri

Alternative Geometries Thought Experiment, 37

pointillist principle

agrees with findings of delayed-choice experiment, 29

an implication of virtual world hypothesis, 17-18

applications in everyday life, 19, 29, 40, 70, 78, 95, 110-111, 131

consistent with the holographic principle, 19

consistent with universal wave function theory, 23

defined, 18, 70

Pring, L, 122

projector of our everyday world

can only move toward, 81, 127

location, 126-127

quantum-level-distributed, 127, 129

slowed by dense objects, 129-131

proof

alternative is experienced truth. 56-57

has meaning only in world of time and space, 50

proprioception, see kinesthetic learning

proving the fundamental ideas of the book, iv

purpose

and the anthropic principle, 140-141

as the elephant in the room, 142

do our lives have, 9, 140-142

purpose of life, primary

and major world religions, 142-144

and our morally evaluative natures, 144-147

and the other people in our lives, 148-150

and the "problem of evil," 147

to choose between good and evil, 142

quantization, of energy, defined, 2

quantum computing

and the human body as a computer program, 64

and the many-worlds interpretation, 119-121

compared with classical computing, 119

described, 119
quantum theory
  and universal wave-function
    theory, 23
  challenge to classical view of
    reality, 2-5
  defined, 2
real time, see time, real
reality
  and quantum theory, 2-5
  classical view of, 2
  dual-reality view of, iii, 9
relativity theory
  and absolute rest frame, 75,
    131-132
religious ceremonies
  and epiphanic experiences,
    156-157
Rogers, Fred, 68
SCAD experiment, 104--106
schizophrenia
  and alternative histories, 65
Schrödinger, Erwin
  espousal of mysticism, 59
script for our everyday world
  where originate, 9, 11, 22, 55, 62,
    70, 79, 97, 126-127, 130, 139
Seattle, Chief, 136 (footnote)

self, increased importance of
  requires new look at life and
    meaning, iv, 136
Socrates
  Allegory of the Cave Thought
    Experiment, 37, 60-61
Soon, Chun Siong, 54
subconscious mind
  as source of creative
    ideas?, 62-64
Tegmark, Max, 30
theory of everything
  and outside-only view of watch,
    Einstein, 9-10
  need for information beyond our
    universe, 27
thinking creatively while in a self-
    induced trance, 33-41-
  answers are discovered not
    remembered, 34-35
  answers from far and near, 37-40
  artists and scientists compared,
    40-41
  by mathematicians, 35, 41
  characteristics of creative thinkers,
    33-34
  compared to creative thinking in
    hypnotic trance, 42, 46

compared to mystical meditation,
31, 32-33

creative thinkers and creative
performers compared, 46-47

process described, 33-34

thinking creatively while in
hypnotically-induced trance,
42-46

compared to creative thinking
while in self-induced trance,
42

compared to thinking creatively in
non-trance state, 43-45

process described, 43

study of, by Dr. Linn Cooper,
43-46

thin places

and epiphanic experiences,
156-157

defined, 156

thought experiment, 36-37

and creation of the physical
concept discovered, 35

and imagery, 37

as part of a new profession, 154

as part of education of scientists,
60

described, 36

importance of role in past
advances of knowledge, 37

time

absolute, 75

and gravity, linkage of, 129-132

and inertia, linkage of, 126-128

an illusion, ii, 153

co-existent with our lives, 137

comparative, see time, comparative

defined, 71, 109

personal, see time, personal

present time as new creation, 28, 75

real, see time, real

truistic definition, 71

time, comparative

and real time, 81-83

by moving in space can only slow,
81, 127

defined, 82

differences in infrequent
circumstances, 85-94

sensation of time slowing, 80, 84-
85, 86, 86-89, 89-90

time, comparative—distortion

during hypnotic trance, 89-94

during impending peril, 86

during peak experience, 86-89

while moving, 81, 84-85, 128, 132

time, comparative—distortion causes
 during hypnotic trance, 98-99
 during impending peril and peak
  experience, 96-97
 during period of sharply focused
  attention, 94-96
 gravitational effect, 80, 83, 95-96,
  97, 98, 99, 104, 128-132,
  132-133
 movement through space, 81, 104,
  127, 128, 132-133
 "watched pot" effect, 94-95, 96-
  97, 99
time, comparative—proving
  distortion
 proof of time distortion possible?,
  103-104
 SCAD experiment, 104-106
 using concurrent reporting in
  hypnosis, 106
 using differences in rate of aging,
  97, 106-109
time distortion, see time,
  comparative—distortion
time, holographic movie model of
 and Bell's theorem, 116-118
 and continual creation, 75-76
 and personal time, 79-81

and the many-worlds
  interpretation, 118--121
 described, 71-75, 100
 implications of, 75, 79
 proving the concept, 100-109
time, personal, 79-99
 illustrated by twin paradox, 79-81
 proving the concept, 103-109
 real and comparative personal
  time, 81-83
time, real
 and comparative time, 81-83
 defined, 82
time reversal violation
 and continual creation, 102-103
 defined, 102
trance, hypnotic
 comparison with self-induced, 42
 induction process, 43
trance, self-induced
 comparison with hypnotic trance, 42
 described, 33-34, 46-47
transcendent experiences of everyday
  life
 and the unitary continuum, 30-31
 compared to mystical experience,
  30-32
 described, 32-33

epiphanic, 156

frequency, 31, 32

objective of, 32, 33

truth

and discovered ideas, 56-58

and thought experimentalists, 56

as an alternative to proof, 56-58

certainty based on experience,
57

discovery by mystics, 56-58

for artists, 41

for mathematicians, 35, 41, 56-57

twin paradox

and real and comparative time,
81-83

and the Aspect experiment, 117

and time dilation for non-living
things, 79-80, 84-85

described, 79

time distortion during well-
established, 104

two-slit experiment, 3-4

challenge to classical view of
reality, 3-4

classic, 3

delayed-choice variation, 27-29

variations of classic, 3-4, 27-29

uncertainty principle

and nondeterministic everyday
world, 78

defined, 2-3

Underhill, Evelyn, 31

challenge to critics of mysticism,
57-58

linkage of mysticism and creative
thinking, 31

unitary continuum

and mystical experience, 30-31

and transcendent moments of
daily life, 31

described, 31

universal wave function theory

and conservation of information,
26, 113

and the many-worlds
interpretation, 23-24

and the pointillist principle, 23

described, 23

universe

as the stage for our lives,
137-138

built like a feedback loop, 28, 75

complete understanding beyond
reach of our science, 9-10

conservation of information, 26,
113

each observer has a separate universe, 24, 141

holographic nature, 18-19, 72, 112-113

life requires fine-tuning of critical variables, 112, 140-141

made of information, not matter, 20

moves through time accompanied by other universes, 110

observer as patch of universe looking at itself, 29, 52

ongoing creation by observer, 28, 75,111

re-created within each frame of holographic movie, 75-76, 102-103

success of inductive reasoning for understanding, 21-22

what is it?, 137

why so vast?, 137-138

universe of the mind (also called world beyond)

and information available to mystics, 14-15, 16-17

as source of projection of script of our everyday lives, 72

creative thoughts and mystical knowledge discovered here, 57, 63-64

defined, 8

knowing, 10-17

looking into, mystics, 11-17

thought experiments well suited for exploring, 37

"watched pot" attentional-focus time-distortion effect

caused by narrow focus of attention, 94-95

described, 94-95

during impending-peril and peak-experience, 96-97

not a factor in hypnotic time distortion, 99

"watched pot" experiment described, 94-95

wave function collapse

and creative thinking, 121-122

and information in other worlds, 26-27

and loss of information, 16, 26

and the many-worlds interpretation, 25-27, 30, 118-119, 120-121, 121-122, 137 caused by human mind only, 138

defined, 25-26

triggered by awareness (attempt
    to measure) particle, 13-14,
    25-26

wave-particle duality

    challenge to classical view of
        reality, 2-5

    defined, 3

waves, particle

    as abstract constructions, 5

    as information only, 5

    what are they?, 5

Wheeler, John, 20, 27-29, 35, 52, 75,
    111, 149

    and present knowledge of both
        past and future, 28

    and the pointillist principle, 29

    delayed-choice experiment, 27-29

information is what makes the
    world, 20

no phenomenon exists until it is
    observed, 28

reality exist because of observing
    the universe, 20

Wigman, Mary, 47

    and kinesthetic learning in dance,
    47-48

    her Pastorale, 47-48

Wigner, Eugene, 21-22

    miraculous appropriateness of
    mathematics for physics,
    21-22

Wolfe, Thomas, 69

world beyond, see universe of the
    mind

Zimbardo, Philip G., 93

ABOUT THE AUTHOR

I bring an outsider's view—hopefully an objective view—to a long-standing enigma for philosophers and physicists alike: What is time?

I am not a physicist. I have a doctorate in an unrelated field from a respected university (University of Pennsylvania), was a long-time member of the Drexel University faculty, and value properly-done research and experimental method. I am an engineer with credentials (BS, West Virginia University) and experience (Armco Steel Corporation).

To me, equally important, I have been, from an early age, a philosopher by avocation with a passion for knowledge, truth, and logical thinking, while also allowing my mind to question given truths and to freely imagine what might lie beyond what we see each day. My interests within philosophy are those of the metaphysicist, the branch of philosophy that deals with the first principles of things, including abstract concepts such as being, knowing, substance, cause, identity, time, and space.

For the past fifteen years, answering the question, "What is time?" has been a consuming interest in my life. I was inspired to continue the process of writing *At the Edge of Time* by the allure of the remarkable truths about our world that unfolded before me as I wrote. I hope reading the book enables you to better see and understand your world as well.

www.ingramcontent.com/pod-product-compliance
Lightning Source LLC
Chambersburg PA
CBHW051456170526
45166CB00001B/274